The Development of Plastics

The Development of Plastics

Edited by

S. T. I. Mossman and P. J. T. Morris

The Science Museum, London

ROYAL
SOCIETY OF
CHEMISTRY

The Proceedings of a Symposium on the History of Synthetic Materials, organized in conjunction with the Plastics Historical Society and held at the Annual Chemical Congress of the Royal Society of Chemistry, Southampton, UK, on 6–7 April 1993.

Special Publication No. 141

ISBN 0-85186-575-5

A catalogue record for this book is available from the British Library

Published by The Royal Society of Chemistry,
Thomas Graham House, The Science Park, Milton Road,
Cambridge CB4 4WF

Printed in Great Britain by Redwood Books Ltd, Trowbridge, Wiltshire

Introduction

When the 1993 Annual Chemical Congress in Southampton was given the general theme of "new materials", the Historical Group of the Royal Society of Chemistry was faced with a quandry: it would be difficult to organise a symposium on the history of new materials! However, it was agreed that Peter Morris would put together a symposium on the history of synthetic materials, and it soon became clear that it would be worthwhile to involve the Plastics Historical Society (PHS) in its organisation. It was fortunate that Peter Morris and Susan Mossman of PHS both worked at the Science Museum. With the help of Colin Williamson (also PHS), we rapidly assembled a programme which covered the history of synthetic materials from the natural and semi-synthetic materials of the Victorian era to the more modern plastics developed during and after the Second World War.

We are grateful that the Royal Society of Chemistry has agreed to publish the proceedings as part of its Special Publication series. The history of plastics (and other synthetic materials) has enjoyed a renaissance in recent years, largely through the efforts of PHS in this country, the tireless activities of the late Raymond Seymour in the United States, and the National Science Foundation-funded Polymer Project of the Beckman Center for the History of Chemistry in Philadelphia between 1984 and 1988. One can detect three major elements in this new wave of polymer history: the questioning of myths, a new emphasis on the cultural aspects of plastics, and greater attention to the economic and political context.

All these features can be found in the present volume. Colin Williamson challenges the assumption that there was no plastic industry before celluloid, and he discusses the working of natural materials such as horn, shellac and bois durci. Susan Mossman analyses the origins of celluloid, a matter which has aroused much debate over the last 150 years, and presents the case for Alexander Parkes' claim to be its inventor. While he acknowledges the pre-eminent role played by Leo Baekeland, Percy Reboul stresses the important contribution made to the development of Bakelite by the British polymath, Sir James Swinburne. Jeffrey Meikle looks at the cultural meanings of synthetic plastics from the perspective of a design historian, alluding to social influences on the changes in and acceptance of plastics as commonplace materials between 1930 and 1945. By contrast, Peter Morris discusses the role of government policy and world wars in the creation of a viable synthetic rubber in Germany and the United States.

The remaining papers, by people who have worked in the field, extend the history of synthetic materials into modern times. Gordon Wilson has worked with polyethylene from its early days, and he relates the early problems with polyethylene (now most commonly known to us in the form of plastic bags). Martin Thatcher discusses the rapid growth of engineering plastics, particularly the polyacetals. Janet Tilley looks at the versatility of acrylics, surveying the early days of Perspex and the many applications of acrylics in the modern world. Brian Parkyn gives a colourful account of the development of fibre reinforced composites from the perspective of one of the early pioneers, who had to solve a variety of problems associated with this important innovation.

The editors would like to thank the speakers for the prompt submission of their manuscripts, Barbara Grant of the Science Museum for retyping the papers to a common format, and Magda Wheatley for her help with the proofreading. We are also indebted to Colin Williamson for his assistance with the organisation of the symposium at Southampton. Finally, we are grateful to Dr Derek Robinson and Dr Robert Bud at the Science Museum for their support.

Susan Mossman and Peter Morris

Contents

List of Figures

Victorian Plastics – Foundations of an Industry

Colin J. Williamson

SMILE PLASTICS, FORD, SHREWSBURY, SHROPSHIRE, UK

The prevalent and frequently stated belief that plastics are a post-1945 phenomenon is as widespread as it is incorrect. Even those currently involved in the polymer industries generally assume that their technologies hardly predate the start of the twentieth century, and whilst this belief could be justified by the introduction of fully synthetic plastics before the First World War, it overlooks the immense amount of research and development into the production of moulded organic polymeric materials in the previous two centuries. Despite these efforts, it is only since the Second World War that the volumes of plastics and synthetic rubber produced annually have become significant in international commercial terms.

Table 1   World Production of Various Materials[1]

Material	Volume of World Production - million tons						
	1913	1938	1950	1960	1970	1980	1989*
Plastics	0.04	0.3	1.5	5.7	27.0	40.0	80
Aluminium	0.7	0.5	1.3	3.6	8.1	11.2	23.1
Zinc	0.8	1.4	1.8	2.4	4.0	4.8	...
Copper	1.0	1.8	2.3	3.7	6.1	8.4	9.4
Pig Iron	53	88	153	241	448	480	570
Rubber-synth	-	0.01	0.5	1.9	4.5	7.7	10.1
Rubber-nat	0.12	0.92	1.9	2.0	2.9	3.7	5.2
Fibres-synth	-	-	0.12	0.65	4.5	8.4	11.6
Cotton	...	5.2	6.0	7.1	7.7	9.1	18
Wool	...	1.6	1.7	2.1	2.2	2.2	2.3

*   Author's estimates
Synthetic fibres excludes semi-synthetic polymers
Most metals indicate primary refined production

From the figures in Table 1 it can be concluded that the commercial breakthrough occurred shortly after the Second World War, but the slow build up of plastics processing as a separate industry can be traced back to the Victorian era and to many of the innovators and industrial entrepreneurs who not only experimented but had clear concepts of the direction of their research, and recognised the importance of new developments as they occurred.

Before exploring the development of the technology we should first define the criteria, which implies that the word "plastics" can be defined, an implication which is difficult to justify. Standard dictionaries define "plastic" by reference to "moulding", but the noun "plastics" is frequently omitted or unsatisfactorily defined. Yarsley & Couzens[2] quote Carleton Ellis, "The term plastic is applied to anything which possesses plasticity, that is, anything which can be deformed under mechanical stress without losing its cohesion, and is able to keep the new form given to it", but this broad definition of the adjective includes many materials that we would not describe as "plastics", and excludes many that we would so describe. Twenty seven years later, in 1968, the same two authors[3] extend their definition to a discussion that admits to being a loose generalisation but which is as precise as is reasonable, "To summarise, a plastic is an organic material......which on application of adequate heat and pressure can be caused to flow and take up a desired shape, which will be retained when the applied heat and pressure are withdrawn."

The 'Oxford English Dictionary' refers to the Greek origins of the word, Pliny and Vitruvius being quoted, with many examples of seventeenth century English usage, which demonstrate the original understanding of the word.[4]

> 1598 R. Haydocke tr. *Lomazzo* 1.7 Painting, Carving and Plasticke are all but one and the same arte.

> 1637 Johnson *Discov., De Progress. Picturae*, The art plastic was moulding in clay, or potters earth anciently.

> 1661 Rust *Origen in Phenix* (1721) I. 75 The beautiful Idea, according to which the Plastick works.

> 1661 Glanvill *Van. Dogm.* To the knowledge of the poorest simple, we must first know its efficient, the manner, and method of its efformation, and the nature if the Plastick.

> 1662 Stillingfl. *Orig. Sacr.* III. i.§ 4 The great enquiry then is, how far this Plastick Power of the understanding, may extend its self in its forming an idea of God.

> 1677 Plot *Oxfordsh.* 251 He (John Dwight) has so far advanced the Art Plastick, that 'tis dubious whether any man since Prometheus have excelled him.

> 1682 H More *Annot. Glanvill's Lux O.* 238 All Souls are indued with the Plastick whether of brutes or men.

"Plastic" therefore referred to a moulding quality both of material and of the souls of men, and clearly the word existed before our modern concept of "plastics". This question of definitions and general perception of plastics is a problem that has been frequently addressed by the twentieth century plastics industry; indeed in the first year of its issue, British Plastics and Moulded Products Trader (the first British journal devoted to plastics), published a letter from C.J. Williams[5] appealing that "The word Plastics should be understood and appreciated as much by the general public as it is by the enthusiasts in our trade." In the following months correspondence flowed with a general conclusion that the word itself was inappropriate, and that better public recognition would be achieved with a new word to fully encompass all the advantages, strength and beauty of the new materials. A "money prize" was proposed for the most acceptable name, many being proposed including:

Korox	Synthoid	Fabron
Chemmold	Chemsynth	Chemfab
Fabric	Mouldics	

and the wonderful "PUCCA", which could be justified by its formation from the initial letters of phenol, urea, cresol, casein and acetate, the "important raw materials used in the plastics industry".

In 1951 the final "s" converting "plastic" to "plastics" was officially acknowledged by the B.S.I. and the B.P.F., and "plastics" is succeeding where "pucca" and the other proposals failed before their launch. Any contemporary definition would probably be based on that of Couzens and Yarsley and would note that paint, adhesives and synthetic textiles are based on similar materials and that silicon based plastics are an anomaly!

If we accept that a concise definition of the materials used is impossible, then an examination of the technologies used might be beneficial. Organic polymers have been moulded for centuries, e.g. beeswax and bitumen, but no industry appears to have been based on the primitive technology tolerated by the low melting point and relatively low viscosity of the melt. The finished "mouldings" were either too brittle or too soft for general applications and when tougher items were required, metals, ceramics, wood, glass and other materials were used. An alternative was horn, a thermoplastic material based on the protein keratin, widely available and extensively used to produce drinking beakers and other containers. The Worshipful Company of Horners is first recorded in the thirteenth century and horn artefacts from this time and earlier are known. Beakers were pieces of oval cross-sectioned horn, straightened over a wooden former after heating, and thin plates of horn were also flattened under pressure for use as transparent sheets in windows and in lanterns (lant-horns).[6] By the 1620s, John Osborne, an Englishman working in Amsterdam, was producing medallions by moulding horn, and by the first decade of the eighteenth century, London had emerged as a capital of the horn moulding industry, producing snuff boxes. We can surmise that the horn was heated either in boiling water or more probably over an open flame in pieces cut to fit the shape of the two piece mould. The mould would be preheated and the horn clamped between the male and female sections using screw pressure. After cooling in the mould, the pressure would be released allowing extraction of the shaped moulding. This would then be trimmed and finished as necessary.

It is possible that moulds were hobbed from a master but the variation in superficially identical designs suggests that several alternative moulds were used, possibly from a softer copper based alloy. Examples of snuff boxes can be found with an indistinct image suggesting wear in use, but the good condition of some of these boxes suggests that the original moulding was made from a worn mould. Probably the best known maker was John Obrisset. His biographer[7] lists about 30 different basic designs, with many variants, mainly showing popular figures, monarchs or classical scenes. Most abundant are snuff boxes depicting the arms of Sir Francis Drake, of which at least five variants are known. That Obrisset and his similar competitors (e.g. Samuel Lambelet) were successful is evident from the number of snuff boxes that survive and when one considers the advantages that he had over his traditional competitors, hand carving their products, it is not unreasonable to assume his success.

Snuff boxes should have reasonably well fitting tops to ensure that the snuff remains in the box when carried in a pocket or bag. Hand carving a top to mate precisely with a base is time consuming, but once a mould is made, then each moulding ejected from it will be substantially identical with the last and will identically mate to a moulded base. Whilst the two halves are being moulded it is no extra effort to mould a design or decoration into one or both halves, thus further differentiating the moulder from the carver in terms of speed of production.

This facility to produce moulded products more quickly and therefore more cheaply than their carved counterpart is the prime motivating force behind the development of plastics and the plastics industry as we know them today.

Whether Obrisset died a wealthy man is not known, but his followers were many, with horn snuff boxes being moulded in many European centres for 150 years. Obrisset probably used cattle horn, but twentieth century horn button makers suggest that buttons and probably medallions are normally made from hoof, which, being less fibrous, is substantially easier to mould. The buttons so made have certainly been described as horn for at least a century, but it is questionable whether pieces thick enough for moulding snuff boxes were available in and prior to Obrisset's time.

Several authors have claimed that the horn was generally powdered before being moulded but there very little evidence to support this theory. Small boxes, especially French music boxes, are available which appear to be from horn or hoof, but which display none of the striations seen in horn nor, on breaking, do they appear to delaminate, and their opacity is greater than normal. It is possible that these items are made from shavings or scrapings of horn or hoof; Hardwick[8] quotes from the *Dictionarium Polygraphicum* of 1735, "mixed with calcined tartar and quicklime and boiled to a pulpy consistency". It is suggested that this is then coloured and cast in metal moulds, but the water content would cause substantial shrinkage on drying and it is doubtful if this was ever a viable method of moulding. It is possible that this method was used to cast sheets which, after drying, were heat moulded as though they were horn, however the mouldings must have exhibited a degree of brittleness without some fibrous content (twentieth century mouldings based on gelatine need wood powder reinforcement). The vast majority of horn snuff boxes that survive today were made from original pieces of horn (or hoof).

By the middle of the nineteenth century three dimensional mouldings were being made, possibly from water buffalo horn, or from the solid tips of European cattle horn. Seal handles can be found moulded as busts of (especially) religious figures with mother-of-pearl inserts and looking very much as though made of a modern thermoplastic material.

In 1861, Prince Albert, husband to Queen Victoria, died and the only jewellery permitted to be worn at court was black, with jet being the major raw material. Jet was expensive to hand carve and hence the high prices commanded by the jet jewellers in and around Whitby attracted many imitators. French jet was black glass, bog oak was carved in Ireland to simulate jet, and in Sheffield there grew an industry moulding horn (and presumably hoof) into brooches. Dyed black with logwood before moulding the brooches are available today in many hundreds of designs. Matching necklaces and bracelets constructed from smaller pieces were produced and the period from about 1870 to 1910 was the heyday of moulded horn jewellery production.

Most horn mouldings found today are still in good condition but those stored in excessive warmth and dampness exhibit delamination, demonstrating the partial release of the stress moulded into the items.[9]

Hoof, horn and the chemically similar tortoiseshell were not the only natural polymers being moulded in the nineteenth century using techniques that the plastics technologist today would recognise. In the Indian sub-continent an industry had developed by harvesting the lac insect (*Tacchardia lacca*) in its larval stage, purifying the resin to produce shellac for the paint and varnish industries of Europe. From a solution in methanol, shellac dries to a hard brittle thermoplastic film, which, on heating or prolonged room temperature storage, cross-links and decreases its solubility in methanol. Its brittleness prevents its use as a potential thermoplastic moulding material, but if reinforced with a fibrous or even particulate filler, shellac can be transformed to a mouldable compound.

In the 1850s photography was gaining popularity, and it was believed that the image in Ambrotypes and Daguerrotypes would fade on exposure to excessive light to leave a mirrored finish on the plate. This has been challenged recently, but the result of the belief was that early images were presented in a folding case, which, if nothing else, allowed a shadow to be cast over the image, thus enhancing its appearance. In England this case was made from wood covered with tooled leather, but in the U.S.A. a composition based on shellac was used. Wood flour was the reinforcing filler and great detail could be reproduced. Over 1000 different designs of these Union Cases are known, from 30 mm diameter circular screw top cases to full plate hinged cases 235 x 190 mm.[10] The ability of shellac to reproduce fine detail justified its continued use until after the Second World War in 78 r.p.m. gramophone records.

Patents granted to Peck, Halvorsen and Critchlow[11,12,13] covered various moulding techniques but curiously not the details of the composition they used. As photographic technology improved, so the necessity for hinged cases decreased and the case manufacturers produced frames and collar box tops. Mineral powders were introduced as fillers to mould mirror & brush backs and brooches. In England many patents were claimed for compositions based on shellac blended with other resins and fillers although few examples exist and it is doubtful if more than a handful ever achieved any degree of commercial success. Exceptions are Manton[14] and Scott with his patented Lionite.[15]

In the vegetable kingdom natural polymers predominate as the structural component of organisms, but lignin and cellulose are not thermoplastic as they decompose on heating. However, resins exuded from trees have been used as film forming materials and thermoplastic mouldable solids, with the major one being from Hevea braziliensis, the rubber tree. The earliest records of rubber being used were in Central America in the sixth century as balls. By the mid-sixteenth century, Spanish explorers were reporting that the Aztecs had a highly developed ball game based on a rubber and in 1615 F. J. de Torquemada reported that Mexicans made shoes, headgear and clothing from the gum obtained from the milk of a tree.[16] C. M. de Condamine introduced it to Europe under the name "caoutchouc" in 1745 and in 1770 J. Priestley introduced the name "rubber" to England as it "erased the marks made by a plumbago pencil".

Natural rubber was dissolved in naphtha (a by-product of the developing coal gas industry) and in the 1820s Thomas Hancock and Charles Macintosh established their company for spread-coating cloth to make a waterproof material. Two layers of canvas were adhered together by the natural rubber thus producing a watertight and airtight fabric. Unfortunately natural rubber is very soft and sticky in hot weather and rigid in cold, a problem which drove Hancock to search for methods to make it more usable. In 1842 William Brockedon is reported to have shown Hancock a sample of treated rubber from the U.S.A. and probably produced by Charles Goodyear. Hancock discovered that sulphur advantageously modifies rubber when the mixture is heated and patented his discovery in 1843.[17] Both Hancock and Goodyear are claimed as the inventors of sulphur modified rubber, but they were both probably predated by Lüdersdorf, Bergius, Leuchs, van Geuns, and others.[18] The addition of sulphur under heat was termed "vulcanisation" by Brockedon, and Hancock later patented the discovery that if up to 30 % of sulphur is added to the rubber, a hard, rigid, thermoplastically mouldable product results. Termed "vulcanite", "ebonite" or "hard rubber", the material was soon to be widely used as an insulating solid by the emerging electrical industry. In the decorative area, vulcanite was moulded to produce brooches, vesta boxes and pipe stems. Good definition was obtainable but as the inherent colour was dark, dull shades were most common, with black in predominance. Vulcanite exhibits surface degradation under exposure to light and moisture, fading to a dull grey-greenish brown with a surface acidity.

Natural rubber, essentially *cis*-1,4-polyisoprene, exhibits elastic properties but its *trans* isomer exhibits plastic properties and is found in nature as gutta percha.   Like rubber, gutta percha is an extract from tropical trees, but unlike rubber, it is hard and flexible at most ambient temperatures, yet softens at approximately 100°C. First imported from the Far East in 1838 it was to become commercially important within 15 years as an insulant for undersea telegraph cables, the first one being laid between England and France in 1850.[19] Bewley developed a screw extruder to coat the copper conductor, thus laying the foundations for the extrusion industry today. Gutta percha was used extensively as a moulding material, the Gutta Percha Company catalogue showing ornamental frames,[20] and was also used as a watertight hosepipe.

In France in 1855, Francois Lepage, "literary man of Paris", patented a mixture of wood flour and albumen as a moulding compound to produce desk accessories and decorative plaques.[21]   His albumen source was egg white or (more likely), blood, and the thermosetting compound produced mouldings of the highest quality. His company, the Bois Durci company exhibited in the Great International Exhibition in London in 1862, and his products are hard to distinguish from highly ornamented, carved rosewood or ebony. As Obrisset had discovered 150 years earlier, making an apparently hand carved item in a fraction of the time it would take to carve had many commercial advantages.

Many other inventors and entrepreneurs produced mouldable compositions in the second half of the nineteenth century, mostly unmarked but some, including carton pierre and papier maché showing a degree of success. The first London International Exhibition of 1851 included 14 exhibitors with gutta percha, 12 with india-rubber, one with lac and three with papier maché, but in 1853 a total of 276 patents were claimed referring to moulding or mouldable compositions. As this is about eight percent of all U.K. patents claimed in that year it is interesting to examine the reasons for this massive interest.

The Industrial Revolution, which started in England, had transformed the manufacture of many products from village craft work to the automated factory established in an urban environment, and made fortunes for the factory owners whose entrepreneurship drove the development forwards. Thus the village weaver saw his livelihood disappear into the mills of Lancashire and Yorkshire; the potteries of Staffordshire and elsewhere took the work of the craft potter and even the village blacksmith was to become merely a shoer of horses and a repairer of the metal products of the Black Country foundries. Simultaneously an increasing number of a growing population was starting to have a disposable income in excess of that required for basic needs of food, warmth and shelter. Deane[22] states of the cotton industry, "The cotton textile industry grew because of a basic requirement for cheap clothing by a growing workforce with a gross domestic product and export markets." Deane further quotes the statistics of the decline of hand looms against power looms over the century.

Table 2 Capital Equipment of the Cotton Industry 1809-1903[23]

	Spindles x 106	Hand Looms x 103	Power Looms x 103
1819-21	7.0	240	14
1829-31	10.0	240	55
1844-46	19.5	60	225
1850	21.0	40	250
1861	30.4	3	400
1870	38.2	---	441
1878	44.2	---	515
1885	44.3	---	561
1890	44.5	---	616
1903	47.9	---	683

The growth rate of U.K. industrial production exceeded 40% per decade from about 1810 to 1850, peaking at over 45% in the 1920s and 1930s. The factory owners controlling this growth became the new wealth creators and, of course, the new wealthy. The growth of the available market was paramount for all industry sectors, and the enlarging home market became the engine for growth for the Industrial Revolution.

"Most industrially successful firms, industries and regions were led by men who saw the potential of a new market, then organised the means of exploiting it, frequently using innovatory technology, and who marshalled the finance necessary to support both production and selling."[24]

These entrepreneurs were constantly searching for new markets, especially as the overall industrial growth rate declined, so the pressure to find new technologies was immense. Few of the original craft skills remained to be automated and an obvious subject of interest was the craft of carving wood, horn, and bone. Combine with this the Victorian love of heavily ornamented decoration, and it is clear that anyone who successfully introduced an automated "carving" technology would be able to build a new industry.

Hence the great number of patent applications from the middle of the century, peaking at nearly ten percent of all those published. A study of the trade mark register from its inception in 1876 indicates that few of these patents were viable, but whether technology, commerce or entrepreneurship was the missing factor is debatable. The moulding techniques had been available since before Obrisset and were essentially the same as those used for stamping metal coins. Extrusion techniques had been developed by Bewley in the first half of the century. Thermoplastic compositions had been laminated onto paper and moulded or embossed and inserts had been used by moulding thermoplastic material around a non-thermoplastic fixed object. Brittle thermoplastic moulding materials had been reinforced with particulate and fibrous fillers.

The market need was there (why else was so much research done?) but all the available compounds were too expensive for anything except small mouldings. What was needed were better materials at lower cost, a search that was to drive Alexander Parkes to discover and launch the first cellulose nitrate based plastic in 1862 (although this was not commercially successful until twenty years later). That same search was not satisfied until Leo Baekeland introduced phenolic moulding materials in the first quarter of the next century, and became the first successful entrepreneur of the plastics industry.

REFERENCES

1.   Source : 'United Nations Yearbook of Statistics'; C. Freeman, 'Economics of Industrial Innovation', 1982; Author's estimates.

2.   V.E. Yarsley and E.G. Couzens, 'Plastics', Pelican Books, Harmondsworth, Middlesex, 1941, p. 10.

3.   E.G. Couzens and V.E. Yarsley, 'Plastics in a Modern World', Penguin, Harmondsworth, Middlesex, 1968, p. 29.

4.   S.A. Simpson & E.S.C. Weiner, 'The Oxford English Dictionary', Clarendon Press, London, 1989. By permission of Oxford University Press.

5.   C.J. Williams, British Plastics, December 1929, I, 254.

6.   P. Hardwick, 'Discovering Horn', Lutterworth Press, Guildford, 1981.

7.   P.A.S. Phillips, 'John Obrisset', Batsford, London, 1931.

8.   P. Hardwick, 'Discovering Horn', Lutterworth Press, Guildford, 1981.

9.   C.J. Williamson, 'Polymers in Conservation', ed. N.S. Allen, M. Edge and C.V. Horie, Royal Society of Chemistry, Cambridge, 1992, p. 2.

10. C. and M. Krainik and C. Walvoord, 'Union Cases', Krainik, Falls Church, Virginia, 1988.

11. S. Peck, U.S. Patent numbers 11758, 1854; 14202, 1856.

12. H. Halvorson, U.S. Patent number 13410, 1855.

13. A.P. Critchlow, U.S. Patent number 15915, 1856.

14. J.S. Manton, British Patent number 2907, 1860.

15. G.A. Scott, British Trade Mark number 9579, 1877.

16. Fordyce Jones, 'History of the Rubber Industry', ed. P. Schidrowitz and T.R. Dawson, Heffer, Cambridge, 1952, pp. 1-9.

17. T. Hancock, British Patent number 9952, 1843.

18. R.W. Lunn, 'History of the Rubber Industry', ed. P. Schidrowitz and T.R. Dawson, Heffer, Cambridge, 1952, pp. 23-27.

19. Telcon, 'The Telcon Story', ed. G.L. Lawford and L.R. Nicholson, Fanfare Press, London, 1950, pp. 29-35.

20. 'Gutta Percha Company Catalogue', 1852.

21. F.C. Lepage, British Patent number 2232, 1855.

22. P. Deane and W.A. Cole, 'British Economic Growth 1688-1959', Cambridge University Press, Cambridge, 1962, p. 191.

23. Source, P. Deane & W. A. Cole, 'British Economic Growth 1688-1959', Cambridge University Press, Cambridge, 1962.

24. E. Pawson, 'The Early Industrial Revolution', 1979.

Parkesine and Celluloid

S. T. I. Mossman

COLLECTIONS DIVISION, THE SCIENCE MUSEUM, EXHIBITION ROAD,
LONDON SW7 2DD

1 INTRODUCTION

In this paper I will discuss the early days of the semi-synthetic plastic which we now know as Celluloid. Several people have laid claim to inventing this material, and my aim is to look at the evidence and attempt to determine the contributions of the various claimants. I will also allude to the early days of the Celluloid industry, with particular reference to the British Xylonite Company, and demonstrate some of the difficulties which the pioneers experienced in making an eventual success of Celluloid.

2 THE DISCOVERY OF CELLULOID

The Background

To begin, it is important to set the discovery of Celluloid in context. The mid-nineteenth century was a period of intense scientific curiosity and invention. People were experimenting with numerous substances ranging from rubber and gutta percha to substances such as shellac and bois durci. Colin Williamson discusses the diversity of materials both natural and semi-synthetic which were around at this time elsewhere in this volume.

Apart from scientific curiosity, much of the impetus behind these new materials came from the need to find satisfactory substitutes for natural materials such as ivory or tortoiseshell which were becoming increasingly rare and/or expensive.

Professor Christian Schönbein, a German-Swiss[1] chemist working in Basle produced cellulose nitrate in 1845. He wrote to Michael Faraday on 27 February, 1846:

> "I have of late...made a little chemical discovery which enables me to change very suddenly, very easily and very cheaply common paper in such a way as to render that substance exceedingly strong and entirely waterproof."[2]

He included specimens with his letter. On 18 March, 1846 he wrote to Faraday again:

> "To give you some idea of what may be made of vegetable fibre I send you a specimen of a transparent substance which I have prepared from common paper. This matter is capable of being shaped into all sorts of things and forms..."[3]

Alexander Parkes

Alexander Parkes, an English inventor, learned of Schönbein's discovery from his agent, John Taylor. Parkes was the ideal man to explore the possibilities of the material further as he already referred to himself as "a modeller, manufacturer, and chemist",[4] and by the end of his life had produced 80 patents in a variety of fields, the majority (66) covering metallurgy but others referring to india-rubber and gutta percha, candles, furnaces, the ornamentation of metals and nitrocellulose.

Fig.1. Alexander Parkes as a young man, by A. Wivell junior, 1848. Courtesy of the Trustees of the Science Museum.

Parkes himself later wrote of this period in his life:

> "I [h]ad but little time to devote to other inventions of the substance
> now known all over the World by the name of Parkesine[.] In the
> year 1852 being more at liberty I began to follow up my Early
> Experiments with a determination to introduce this most interesting
> Invention and for which I [h]ad a devoted love - so much so that I
> gave nearly 5 years of my Extra time to Chemical Experiments
> assessed by my brother Henry Parkes whos[e] Chemical Knoledg[e]
> was allwas of great value to me Espethr in Preparing the Nitro
> Selose which I [had made] from all kind of substances such as the
> finest Paper [and] Coton Materials." [5]

To nitrate his "coton", Parkes used a mixture of nitric and sulphuric acid. He
mixed these constituents with vegetable oils and small proportions of organic
solvents producing a mouldable dough which he christened Parkesine. Objects were
produced from the dough, which was heat-softened and then pressed into moulds;
alternatively they were hand-carved and then inlaid with mother-of-pearl or metal
wire. Parkesine was displayed in a variety of colours and forms at the Great
International Exhibition of 1862, some of which are held in the Science Museum's
collection.[6] A guide to the 1862 Exhibition observes that:

> "Among the most extraordinary substances shown is a new material
> called 'Parkesine' from the name of its discoverer, ... a substance
> hard as horn, but as flexible as leather, capable of being cast or
> stamped, painted, dyed or carved and which above all can be
> produced in any quantity at a lower price than gutta percha."

Parkes himself noted:

> "I Produced A great variety of articles so early as 1860. and at the
> Great English Exhibition of May 1.1862. I Exhibit[ed] the invention
> for the first time to the world in my Case Class 4 Chemical section
> ..."

including

> " Combes Book Cover Billiard Balls Boxes Enlayed with Pearl
> Silver & Gold Pen's. Telegrap Wire woven fabrics Covered with
> Plain and Colored Parkesine, Buttons, Knife handls, Paper knives.
> Medallion Gumes with artificial teath for dental Purposes imitation
> tortoiseshell- and malechite Japelow? venered workes. Tubs?
> musical instruments Pinons? Keys Surgical Instruments- Specimen?
> Bosses drinking vessels Brush Backs and Transparent Sheets
> Colourless & of all Coulours. I by this time [h]add many Patent[s]
> in England & other Countrys - and now I wish to point out to you
> what I did to make all these articl[e]s? of Niotro Celulois &
> afterward Caled Parkesine."[7]

Fig.2. Selection of Parkesine objects. Courtesy of the Trustees of the Science Museum.

He received a bronze medal for excellence of quality, and a silver medal when he exhibited his discovery at the Paris Universal Exhibition in 1867 (both these medals are now in the Science Museum Collections).

In 1865, at the Royal Society of Arts, Parkes stated that:

"Another important improvement in the manufacture of parkesine is the employment of camphor which exercises an advantageous influence on the dissolved pyroxyline, and renders it possible to make sheets... with greater facility and more uniform texture, as it controls the contractile properties of the dissoved pyroxyline; camphor is used in varying proportions according to requirement, from 2% to 20%."[8]

This knowledge of the useful properties of camphor is an important point and will be referred to again.

Parkes decided to market his invention and launched the Parkesine Company Ltd in April 1866, with a capital of £10,000, some of which was his own money. However, following initial optimism, the company failed to make a commercial success and went into liquidation in 1868. Failure of the Parkesine Company has been blamed on both on the flammability of Parkesine, and on the use by Parkes of inferior materials. Indeed this is indicated by Parkes himself in a later document where he said that:

"from the aims of the Company to Produce the Cheapest Possibl NitroSelulus for Parkesine i[t] was quite unnecessary to use the fine Colour or Papers I used at first and only the Cheapest and commercial - materials ... some so low in Price as 1/-? lbs."[9]

This might explain some of the complaints that his:

"combs sent out in a few weeks became so wrinkled and contorted as to be useless."[10]

These symptoms have been explained by Worden as being the result of the Parkesine Company producing too much Parkesine too quickly, before all the solvents could be eliminated.

Until recently it has been accepted that Parkes ceased working on cellulose nitrate after the failure of his company; however evidence from his unpublished notebooks and letters in the possession of the Plastics Historical Society reveal that in 1881 he returned to the problem of nitrating cellulose.

"Since my old patents have all expired I have made great improvements in Cellulose and Celluloid and especially in preventing the rapid combustion of Cellulose substance and Nitro-Cellulose such as Xylonite or Celluloid. My new inventions enable me to make an uninflammable Cellulos or Celluloid and will cause this beautiful substance to be imployed in a much larger manufacture than before this time, and give greater confidence in the Artisans employed in the manufacture of articles of Celluloid and to the public all over the World. My brother Henry Parkes has also taken two patents recently for improvement in Celluloid and a company is now formed and was registered on the 26th of Feby 1881 under the name of the London Celluloid Company...which will be established at Crayford in Kent where every variety of articles will be manufactured of uninflammable Cellulose and Celluloid." [11]

Unfortunately such enthusiasm was unfounded and the London Celluloid Company collapsed very rapidly.[12]

Why had Parkes not worked on Celluloid until 1881? Patent assignments held in the Hackney Archives Department reveal that Parkes assigned his patents to the Xylonite Company in 1869 (with the permission of the liquidator of the Parkesine Company, Charles Augustus Harrison), and had undertaken that he would:

" ... not at anytime hereafter during the continuance of the ...Letters Patent ... be engaged as principle partner Agent Manager or Workman in any Manufacture...of compounds or materials containing Xylodine [Cellulose Nitrate]... will not in any way trade or deal in the said substances & materials."[13]

In addition he had undertaken that if he discovered any improvement in the manufacture of Parkesine and related substances:

> "...he will at all times hereinafter without any further consideration than aforesaid communicate [these] to the said Xylonite Company."[14]

Another document dated 1873 [15] assigns to Daniel Spill:

> "All those the Letters Patent & Patent rights under the Great Seal of Great Britain dated the 8th day of December 1865 ... granted to the said Alexander Parkes ....." concerning "...his said Invention of Improvements in preparing compounds of Xylodine or Gun Cotton & in the apparatus employed in that part of the United Kingdom of Gt Britain & Ireland called 'England' from the day of the date of the said Letters Patent for the term of 14 years thence."

(ie: until December, 1879). It was only when Parkes' patents and hence his assignments to the Xylonite Company (which were later assigned to Daniel Spill) expired, that he returned to the problem of nitrating cellulose in 1881. Evidence reveals that Parkes not only continued his work on cellulose nitrate after a break following the collapse of his company in 1868, but that he was also fully aware of the importance of camphor as a solvent for celluloid. In September 1881 an entry records another recipe for solvents and camphor:

> "by this addition the bulk is increased to 7 gallons when camphor goes in to increase the bulk of this celluloid."[16]

Daniel Spill

In the meantime, Daniel Spill, who had been Parkes' Works manager, tried to make money out of Parkes' invention which he renamed Xylonite, somewhat to Parkes displeasure. Spill set up the Xylonite Company in 1869 and produced items made of Xylonite and Ivoride ranging from hand mirrors and fancy combs to knife handles, as well as curiosities such as a death's head walking stick handle now in the possession of the Science Museum and which was Spill's own. Spill's price list for October 1869 calls Xylonite:

> "An excellent substitute for ivory, bone, tortoiseshell, Horn, Hard Woods, Vulcanite etc. - it is not at all affected by chemicals or atmospheric changes, and therefore valuable for shipment to hot climates."[17]

Despite these optimistic claims for his product, Xylonite had great limitations due to its flammability, although it was relatively easy to produce, hence Spill met with little success in his business venture.

The Hyatt Brothers

An American, John Wesley Hyatt was also experimenting with cellulose nitrate and experiencing the problems with flammability already suffered by Parkes and Spill.

Hyatt later commented as follows about the manufacture of billiard balls by his Albany Billiard Ball Company:

> "In order to secure strength and beauty only colouring pigments were added, in the least quantity. The result being that the billiard balls were coated with a film of almost pure gun-cotton. Consequently a lighted cigar applied would at once result in a serious flame and occasionally the violent contact of the balls would produce a mild explosion like a percussion guncap. We had a letter from a billiard saloon proprietor in Colorado, mentioning this fact and saying that he did not care so much about it but that instantly every man in the room pulled a gun."[18]

A more stable product was needed. Hyatt made the break-through in 1870 when he patented the fact that camphor makes an excellent solvent and plasticiser for cellulose nitrate.[19] He called his product Celluloid, and this is the name which is most commonly used today for products based on cellulose nitrate. By 1872 the "Celluloid Manufacturing Company" was in existence and became an enormous commercial success for the Hyatt brothers, producing cheap popular goods, and in particular combs, collars and cuffs.

Spill was most unhappy with the Hyatts, claiming that they had infringed his patented processes. Lengthly litigation resulted, lasting from 1875 to 1890 in which Parkes gave evidence for the Hyatts. Spill lost the battle and returned to England a ruined man, soon dying of diabetes in 1887[20], his death finally resulting in the dismissal of the case by the U.S. Supreme Court.[21]

In the meantime, Hyatt had made significant advances in the machinery used for the production of celluloid. He had already benefitted from those developed for natural plastics, namely Samuel Peck's screw press with its steam-heated platens, Hancock's masticator, Bewley's extruder and Real's hydraulic press of 1816. He also used Hancock's slicer of 1840 to make celluloid sheets by slicing thin sheets from large moulded blocks. Hyatt then collaborated with a practical engineer by the name of Charles Burroughs in the design and construction of specific machinery for manufacturing his material.[22] Burroughs designed specialised tools and machinery for celluloid production including the "stuffing machine" which produced celluloid "in the form of a bar, sheet or stick" which was then machine-finished.[23]   This is considered to be the precursor of the modern injection moulding machine. Burroughs also invented the compression sheet moulding press and the hydraulic planer which sliced celluloid slugs into thin sheets. "Burroughs Blowing Press" was also conceived by Hyatt; this machine made celluloid tubes or sheets expand to fit the contour of the mould: an early example of blow-moulding.

Hyatt described his achievements as follows:

> "First the idea of combining with the nitrocellulose only the exact or approximate amount of solvent required for a solid solution. This required a nearly perfect mechanical mixture before very much solvent action could take place. Second, completing the solution by means of heat and pressure. Third, eliminating the practically unnitrated fibres by pulping. Fourth, employing tissue paper in place of carded fibre. Fifth, avoiding the terrible danger of drying out the

moisture by exposure in a 'drying room'...Sixth, the stuffing machine process. Seventh, the sheeting process, most essential."[24]

Celluloid was used for a range of objects from the 1870s onwards, including articles of apparel such as collars and cuffs, as mentioned above. It was also widely used to make cosmetic boxes which appeared in a variety of finishes ranging from plain to luxurious imitation Loetz iridescent glass (based on an Art Nouveau Parisian glass) and was often made to imitate natural materials such as ivory, mother-of pearl, shagreen and tortoiseshell. One design for tortoiseshell was made by Colin Glover, a descendant of the Merriam Family, the founders of the British Xylonite factory. The effect was produced by interleaving different coloured sheets of celluloid, compressing, slicing, then interleaving again, and reslicing four times in total to produce a tortoiseshell effect.[25] Glover was confident that the British Xylonite Company could copy any pattern presented to them in celluloid. Such designs were not patented but recorded in the company's configuration book with a sample, signed and dated.[26] Celluloid can also boast its own masterpieces, such as the elaborate combs made by **Auguste Bonaz** and Clement Joyard collection, and housed in the Museé des Peignes, Oyonnax, in France.

The flammability of Celluloid remained a problem. One solution was to substitute the related material, cellulose acetate, which had a low combustibility and this was first produced on an industrial scale in the form of dope which was used to coat the fabric wings of aeroplanes during the First World War. Cellulose acetate replaced the dangerously flammable cellulose lacquers, but was not produced in a mouldable powder form until 1929, and so cannot be thought of as a widely used early plastic. Indeed, those who worked in the industry considered celluloid to be a far more beautiful material.[27]

3 THE ORIGINS OF CELLULOID: A DISCUSSION

Unpublished notebooks and papers of Alexander Parkes (in the possession of the Plastics Historical Society) have helped to throw some light on the origins of celluloid.

Parkes speaks from the heart in this letter:[28]

> "Sutton Coldfield March 7th 1881
> "...I trust I may be at last rewarded for more than 30 years labour which I have devoted to these inventions of Parkesine Xylonite, or Celluloid, and I may also add, that I have taken some 20 patents for various improvements and applications of this wonderful substance and the money value it has cost me apart from the time devoted to it would be considered by many a fortune, which I have spent freely from a love and desire to perfect what I began under very difficult circumstances, and I have the pleasure if not the <u>profits</u>, of knowing that in England, America, and France large works have been successfully established.
> Alexander Parkes"

It is worth examining further the background to the development of Celluloid, looking in more detail at the three main proponents of its invention: Alexander Parkes, Daniel Spill and John Wesley Hyatt.

A note of Alexander Parkes dated to c. 1881 indicates that he thought very little of Spill's claims to originality:

> "The Nitro Selulose was made afterwardes and Bleached in accordance with my instruction .... when the vessel ware op[e]ned by me in the Presence of Spill... I well Remember Spill Expressing supp[r]ise when he pulled out a handfull of the Nitro Selulous at finding of so Perfectly while & apparantly unchanged - he said what the Devwl is it, I felt by his Remarks quite sure he was Ignorant at this time that he did not know what Nitro Selos was like as it was, but little altered in appearance from the original Cotton by the Chemical Reaction in Changing it into Nitro Selulose - I actually Explained to him how it was made and Bleached after it was made ....I think [I] need not say any more on the Preparation of NitroSelulose.... only this as there is no Invention in Bleaching Nitro Selulose there could be not any Patent Right[.]"[29]

In the same document Parkes stated that:

> " I particularly wish to show you Clearly I was using Camphor and combining of Camphor as a solvent of NitroSelulose long befor I became encumbered? with D. Spill - I first   discovered that Camphors & Camphor with Alcohol was a solvent of Nitro Selulos ....in the year 1853 .... by putting Camphor in a test tube with Nitro Selulos and heating it slowly at a low heat for a short time or at a lower temperature? for a few hours when the Cotton will appear in the test tube unchanged but on taking it out and Pressing it you find that the Solvant action [h]as taken Place for the Cotton Celuloid Become a Part and setts as hards as Ivory afterwards[.] From this facts I Patent the Imployment in the Manufacture of Parkesine which become like the Solvant Alcohall the Property of the Parkesine Camphor...in my Printed English Patent I wish to inform you that after the Parkesine Company was disolved I sold my various Patent[s] to other People for whom Daniel Spill becom Manager of not a Partner with them and that agreement is still Running to the End of one of my Patents when I shall force to Extend the tenor of Patent if I find it desirable and no other person can do this."

Not only does this document firmly declare that Parkes knew the value of camphor as a solvent for Celluloid, it also gives a less than flattering view of Spill, highlighting his appearance on the scene after Parkes had already begun to combine camphor and nitrocellulose. The truth of Parkes' statement that he did not sell his patents to Spill but to a company for which Spill worked is borne out by the contents of the patent assignments made by Parkes to the Xylonite Company in 1869.[30] Spill himself was not assigned the patents until 1873.

What of Hyatt and his claim to be the inventor of celluloid? Hyatt himself stated in 1885 [31] that:

"We are aware that pyroxyline has been heretofore subjected to the action of spirits of camphor or other solvents, and do not, therefore, broadly claim such process, but to our best belief and knowledge no successful means or apparatus have until now been devised to accomplish satisfactory results in economically, uniformly, and thoroughly mixing the materials."

This neatly summarises what I consider to be Hyatt's contribution to the history of celluloid. He refined the production of a substance which Parkes had invented, devising various ingenious means by which the nitrocellulose might become impregnated with camphor. He was also clever enough to patent the role of camphor:

"The principal feature of the process ...consists in employing camphor gum as a solvent of pyroxylin pulp."[32]

Hyatt then proceeded to market Celluloid successfully by making quantities of popular items.

The origins of celluloid is a matter of continuing debate amongst various scholars of plastics history. The American, Robert Friedel, who has made a detailed study of the history of celluloid, favours John Hyatt as the inventor of a marketable cellulosic product. Indeed he says Parkes' company failed due to the overuse of cheap raw materials in an aim to keep the product as inexpensive as possible;[33] the British scholar, the late Morris Kaufman, was of the same view.[34] Colin Williamson supports Alexander Parkes.[35] The argument hinges around camphor. Did Parkes recognise the value of this material as a solvent as well as a plasticiser for cellulosic plastics?

Gretchen Shearer (formerly with the Institute of Archaeology in London) analysed some samples of Parkesine from the Science Museum Collections and has commented that they contained very high proportions of camphor.[36] They are also, to my own knowledge, in excellent condition and are dated to between 1855-1868. If this is an accurate date, then that would seem to prove that Parkes did know the value of camphor. In his memoranda of 1879 and letters of 1881 he states this quite clearly. He also referred to this as early as 1865 in his lecture to the Royal Society of Arts:

"Another important improvement in the manufacture of parkesine is the employment of camphor."[37]

Why else would Parkes have given evidence against Spill if not because he felt he himself had a better claim? He makes this clear in his notes, as well as in his sworn evidence before a U.S. commissioner in 1878 when he stated :

"It is true that I used alcohol and camphor at that early date [1853]...I was the first to discover the fact that alcohol alone was a solvent of nitrocellulose and I published that fact in my patent of the year 1855 and I also soon after discovered the fact that camphor alone was a solvent of nitrocellulose."[38]

These words were spoken under oath. Does one believe Parkes? He, by all accounts, was a moral man, charitable to the unfortunate and apparently not overly self-seeking. Otherwise he would not have refused to give the secret of a powerful explosive to the Government (at the request of his wife). He also showed patriotism in declining foreign offers for his explosive invention.[39] Moreover, one should not undervalue the fact that Parkes gave sworn evidence as he seems to have been a religious man, judging from the motifs he used for some of his Parkesine objects: Madonnas as well as a miniature head of Jesus Christ, and one of his first uses of Parkesine was to make Bible covers.

Compare his character to Daniel Spill's, a man who had given up a medical doctor's career for a life in business, and was described by Charles P. Merriam, the son of one of the founders of British Xylonite, as:

> "a great duffer..." who "...had'nt any notion how to make anything good."[40]

Merriam has also stated that:

> "Spill resigned from the board...went to America and and started negotiating the sale of American patents he had already sold to us."

(ie: the British Xylonite Company) and pointed out that

> "Spill appears to have acquired the patents."

Spill's exact status is unclear and one begins to wonder about his legal right to these patents (presumably most taken out by Parkes), particularly as Parkes himself wrote:

> "I wish to inform you that after the Parkesine Company was disolved I sold my various Patent[s] to other People for whom Daniel Spill becom Manager of not a Partner with them and that agreement is still Running to the End of one of my Patents when I shall force to Extend the tenor of Patent if I find it desirable and no other person can do this."[41]

In fact Spill did not acquire the patents himself until 1873[42]. Dr John Goldsmith who at one time worked as a chemist for British Xylonite, carried out analyses of Parkesine in 1922. In his letter conveying these results, he noted the presence of camphor and also stated:

> "he (Parkes) could not refute Spill's claims to be the inventor of camphor-celluloid by referring to these...samples without risking the validity of his patents."[43]

This is supported by the following note by Parkes, dated to 1881 in which he stated:[44]

"After the Parkesine Company closed at hackney Wick a new company was formed called the Xylonite company, of which company D. Spill (late works manager of the Parkesine company) was one, and the original name Parkesine was suppressed and the name Xylodine was given to the same substance. A Parkes having the patent rights still in his possession granted to the new company an absolute license and sale of five of his patents, and agreed with them not to work himself, in any manufacture of Parkesine, Xylonite, or in any manufacture where Nitro-Cellulose was employed, during the continuance of either of A. Parkes' patents, all of which have for some time passed expired, and by this arrangement the new company continued the manufacture of what they 'now call Xylonite until the present time with more or less success, so that his original inventions never ceased to be worked in England. I believe that the American Celluloid company have greatly extended and forced this manufacture, and by their spirit and Ability have created a large and Profitable business in the United States, but this manufacture only commenced long after I established the original works in London about the year 1864 or 5…I must also say that my first patent was granted to me in the year 1855, and long before I found by experimenting in my laboratory that Alcohol and Camphor were solvents of Nitro-Cellulose without addition and yet this fact even up to this time is still denied by one patentee [ie: Spill] in a recent trial in the United States. But in one of my patents of 1864 and one of 1865 I claimed the use of Camphor for certain advantages, and my use of Camphor was largely Employed in England years before D. Spills Patent of December 31-1868 or before the Brother Hyatt's Patents were granted in America as the dates of the two patents of mine referred to above and granted to me in 1864 and 1865 will shew…"

Parkes was clearly concerned that his contribution to the history of celluloid had been forgotten, and wanted to make it obvious that he knew of the importance of camphor in relation to nitrocellulose at least as early as 1864, well before Spill and the Hyatt brothers made a claim on the process.

In conclusion, Alexander Parkes wrote a letter in 1881:[43]

"Sutton Coldfield
March 7th 1881

G Lindsey.
25 Bernells Hill <u>Birmingh</u>

Dear Sir   In answer to the American Inquiry "Who Invented Celluloid"
                I have put together a brief history of my various patents for the invention of Parkesine, Xylonite, or Celluloid for they are all the same. I do wish the World to know who the inventor really was, for it is a poor reward after all I have done to be denied the merit of the invention and you will be able if you make any use of this information to place the inquiry as to who invented Celluloid in a true and just form before the readers of your paper ....
                Yours very truly
                                Alexr. Parkes"

The aim of this paper has been to try to redress the balance. I hope that this had been achieved, and that the reader is persuaded, once and for all, that Parkes truly was the inventor of Celluloid and the Father of the Plastics Industry.

ACKNOWLEDGEMENTS

I am grateful to the Plastics Historical Society for allowing me access to the unpublished notebooks and papers of Alexander Parkes.

REFERENCES

1.   Schönbein was born in Swabia, Germany, but worked in Switzerland.

2.   G. W. A. Kahlbaum & F. V. Darbishire, 'The Letters of Faraday and Schoenbein', Williams & Norgate, London, 1899, pp. 152-153.

3.   G. W. A. Kahlbaum & F. V. Darbishire, 'The Letters of Faraday and Schoenbein', Williams & Norgate, London, 1899, p. 155.

4.   R. D. Friedel, PhD Thesis, 'Men, Materials, and Ideas: A History of Celluloid', The John Hopkins University, 1976, 21.

5.   Plastics Historical Society Archive, undated note by A. Parkes, Inverness (c. 1881).

6.   Science Museum Inventory number 1937-30.

7.   Plastics Historical Society Archive, undated note by A. Parkes, Inverness, c.1881.

8.   A. Parkes, 'On the properties of Parkesine and its Application to the Arts and Manufactures', Journal of the Society of Arts, December 22, 1865, 14, 83.

9.   Plastics Historical Society Archive, undated note by Parkes, Inverness, c. 1881.

10. E. C. Worden, 'Nitrocellulose Industry', D. Van Nostrand, New York, 1911, pp. 570-71, note 2.

11. Plastics Historical Society Archive, E10, p.3.

12. J. N. Goldsmith, 'Alexander Parkes, Parkesine, Xylonite and Celluloid', 1934, p. 54. Bound manuscript, Science Museum Library and Birmingham Reference Library.

13. Hackney Archives Department, no.D/B/XYL/14/1: Alexander Parkes Esq and Another to the Xylonite Co. Ltd - Assignment of Letters Patent. 23 June 1869.

14. Hackney Archives Department, no.D/B/XYL/14/1: Alexander Parkes Esq and Another to the Xylonite Co. Ltd - Assignment of Letters Patent. 23 June 1869.

15. Hackney Archives Department: D/B/XYL/14/2, Assignment of Letters Patent/ The Liquidation of the Xylonite Co. Ltd & the said Company to Mr Daniel Spill 8 Dec 1873.

16. C. Williamson, Plastiquarian. Spring, 1989, 2, 8.

17. Science Museum Technical File, 1937-30.

18. J. W. Hyatt, 'Address of Acceptance' [of the Perkin Medal], Journal of Industrial and Engineering Chemistry, 1914, 6, 158-61.

19. J. W. Hyatt & I. S. Hyatt, U. S. Patent , 12 July, 1870, Improvement in treating and molding pyroxyline.

20. Science Museum Library, Kaufman Archive, Interview with Mrs Frith (Spill's grand daughter), 1962.

21. R. Friedel, 'Pioneer Plastic', The University of Wisconsin Press, Madison, 1983, p.131.

22. J. H. du Bois, 'Plastics History U.S.A.', Cahners Books, Boston, 1972, p. 42.

23. U.S. Patent 133,229, 1872.

24. J. W. Hyatt, 'Address of Acceptance' [of the Perkin Medal], <u>Journal of Industrial and Engineering Chemistry</u>, 1914, <u>6</u>.

25. Glover (op.cit.) referred to this particular design as no.1705.

26. Colin Glover, Interview by P. Reboul, 29 September, 1981; many of these configuration books are now in the Ipswich Public Records Office.

27. Colin Glover, interview, 29 September, 1981.

28. Plastics Historical Society Archive: E10, paper number 4.

29. Plastics Historical Society Archive: undated note by Parkes, Inverness, c. 1881.

30. See no. 13.

31. J. W. Hyatt & J. Everding, U.S. Patent 326,119, September 15, 1885: <u>Process of making solid compounds from soluble Nitrocellulose</u>.

32. J. W. Hyatt & I. S. Hyatt, U.S. Patent 133,229 , November 19, 1872: <u>Improvement in Process and Apparatus for manufacturing Pyroxyline</u>.

33. R. D. Friedel, PhD Thesis, 'Men, Materials, and ideas: A History of Celluloid', The John Hopkins University, 1976, 25, 28.

34. M. Kaufman, 'The first Century of Plastics', Plastics Institute, London, 1963, p. 24.

35. C. Williamson, <u>Plastiquarian</u>, Spring, 1989, <u>2</u>, 8.

36. <u>Modern Organic Materials Meeting</u>, The Scottish Society for conservation & Restoration, Edinburgh, 1988.

37. A. Parkes, 'On the properties of Parkesine and its Application to the Arts and Manufactures,' <u>Journal of the Society of Arts</u>, December 22, 1865,<u>14</u>, 83.

38. R. D. Friedel, PhD Thesis, 'Men, Materials, and ideas: A History of Celluloid', The John Hopkins University, 1976, p. 27ff.

39. J. N. Goldsmith, 'Alexander Parkes, Parkesine, Xylonite and Celluloid', 1934, p.37. Bound manuscript, Science Museum Library and Birmingham Reference Library.

40. C.P. Merriam, Private Journal, BXL Archive no.6/2, Science Museum Library.

41. Plastics Historical Society Archive, undated note by A. Parkes, c.1881, Inverness, p. 10.

42.   Hackney Archives Department: D/B/XYL/14/2: <u>Assignment of Letters Patent/ The Liquidation of the Xylonite Co. Ltd & the said Company to Mr Daniel Spill 8 Dec 1873</u>.

43.   Science Museum Library, Kaufman Archive.

44.   Plastics Historical Society Archive, E10, Paper Number 1.

45.   Plastics Historical Society Archive: E10.

Britain and the Bakelite Revolution

Percy Reboul

PLASTICS HISTORICAL SOCIETY, I I HOBART PLACE, LONDON SWIW OHL

The discovery and development of new products and processes is only rarely the happy accident or blinding flash of inspiration so beloved of the tabloids and other writers of fiction. It is much more likely to be the result of steady, persistent effort based on or guided by earlier workers in the field. It also sometimes involves different people working independently on the same problem in different parts of the world. This is certainly true of the development and commercial exploitation of phenol-formaldehyde resins, popularly called BAKELITE.

However, before looking at the historic details (particularly the British experience) it is important that we remind ourselves briefly of those years leading up to and just after the turn of the century. Scientific thought and invention was changing the world profoundly. What an age it was ... wireless telegraphy, the internal combustion engine, aviation, electricity, iron ships, railways, photography ... the list is immense.

One barrier to progress, however, was that natural raw materials could not always provide the properties and consistency of quality necessary to meet the demands of the emerging new technologies. Almost as important was the need for sources of raw materials that could be better controlled in market terms and was not subject to the vagaries of climate or political upheaval. Ever present, too, was the need to contain costs and improve profitability. A classic example of this scenario was the successful development of synthetic dyestuffs.

By way of a specific example, we may quote the remarks made by Sir James Swinburne FRS, who was the first President of the Institution of Electrical Engineers and, as we shall see, one of the great pioneers of the plastics industry. In his Presidential address in 1902 he said:

> "It is largely the insulation of cables that limit our pressures and therefore our distance of transmission. The conductor itself (ie copper) can hardly be improved upon but there is room for great improvement in the insulation".

The insulants in this case were rubber and gutta percha and it is hardly surprising, therefore, that scientists such as Swinburne were ever alert for new materials with improved properties. Plastics were to provide many of the answers.

The development of phenolics is essentially the story of two remarkable men: Sir James Swinburne, a British baronet who lived to be 100 years old, and Leo Hendrik Baekeland, a Belgian chemist who settled in the U.S.A. Both men were interested in what to most people of the time must have appeared the strange phenomenon which occurred when phenol and formaldehyde were reacted together - namely a sticky, amber-coloured resin, rather like treacle, which hardened upon cooling and had a number of very desirable properties. Others before them had made a study of the reaction: Adolf von Baeyer, Werner Kleeberg and Adolf Luft among others. Baekeland and Swinburne, working independently, were seeking a practical method of mass producing the resin, controlling its latent violence and harnessing its properties in a form and at a price that might attract some of the emerging industries which were referred to earlier. The paths of the two men were to cross and re-cross over the years but what finally emerged was a product and an organisation that was to change the world. It has to be said, right at the start, that Baekeland's work was a masterpiece of systematic investigation and far superior to Swinburne's researches. He is rightly regarded as "the father of the plastics industry".

However, Swinburne is an equally fascinating character. Both he and the British experience are much less well known than their American counterparts but are a remarkable story of persistence and triumph over seemingly impossible odds.

Leo Hendrick Baekeland

Leo Baekeland was born in 1863 in Ghent and showed early signs of academic promise. At evening classes he won gold medals in each of his four subjects: chemistry, physics, mechanics and economics. We may perhaps reflect on that latter subject because it was his flair for business that made him such an outstanding success when it came to commercially developing his inventions.

He entered the University of Ghent in 1880 and it took him only 2 years to achieve his Bachelors degree and only 4 for his Doctorate. At the remarkably early age of 26 he became a Professor at the University and was soon offered a travelling scholarship that was to change his life.

A visit to British universities did not impress him - he felt their attitude to science was all wrong. In the U.S.A., however, he liked what he saw and was persuaded to work and settle there, opting for a commercial career rather than an academic one. He became a consultant and it was not long before he perfected and patented Velox photographic paper which could be printed and processed far more easily than the conventional photographic papers of the time. He was soon to sell the rights to Eastman Kodak who paid him a vast sum even for that time - $750,000 - which made him financially independent and able to pursue his other scientific interests, one of which was the phenol-formaldehyde reaction.

His work on the preparation and isolation of the resinous product, obtained by the condensation of phenol and formaldehyde, was a masterpiece of scientific investigation and his patents, notably one taken out in 1909, became the cornerstone of the industry.

His patents show that he had virtually complete understanding of the reaction and could control it. Among other things he found methods by which the resin could be mixed with sawdust and colorants to produce a moulding powder which, when subjected in a mould to heat and pressure, would produce a solid, infusible, insoluble moulded shape with excellent mechanical and electrical properties. He called his material BAKELITE and by the end of the first decade of the 20th century had established a firm base in the U.S.A. for manufacture and applications of phenol-formaldehyde resins, had taken out patents in the U.S.A., Germany and France, and formed companies in all three countries to exploit them.

Swinburne

Sir James Swinburne Bart, F.R.S., was born in Inverness in 1858. His father was a Royal Navy Captain and he was educated at Clifton College where his tutors imbued in him a love of the natural sciences and engineering. At an early age he became an outstanding figure in the electrical industry.

Fig.3. Sir James Swinburne. Courtesy of the Trustees of the Science Museum.

Among the accomplishments noted in his application for Fellowship of the Royal Society (he was nominated, incidentally, by some of the most distinguished scientists of the day including Kelvin, Oliver Lodge and William Crookes) was

> "intimately connected with the development of the dynamo and the theory of electrical measuring instruments."[1]

Incidentally, the words "stator" and "rotor" (as in rotor arm) were coined by him and he worked with Swan on the development of the first electric light bulbs.

He had an international reputation as an expert witness in patent disputes; was an authority on the metallurgy of razor blades; a scientific consultant to General Motors; and was the author of books on music, clocks and watch movements.

We can only wonder at the skills and capacity for work of these Victorian polymaths. When you combine such talents with a gift for poetry, personal charm and a good sense of humour, you have a person exceptional in every sense of the word.

Here, in Swinburne's own words, is the moment in time when he first became aware of the product of the phenol-formaldehyde reaction:

> "Sometime about 1902, I went into the office of Hermann Voltereck. He was a patent agent by profession, but he was also an inventor of a wild character. On his table was a lump that looked as if half a pint of beer had frozen and come out of its glass. Voltereck saw no future in this but I said there was a future. We formed a little company (in 1904) called The Fireproof Celluloid Syndicate Limited. We got Luft, the inventor over from Austria. He had patented the reaction. He was a chemist in a scent works who was working to see if he could get a disinfectant and had produced this body and various articles made of his material which was tough and somewhat flexible. The process had serious drawbacks and we therefore set to work in my laboratory in 82 Victoria Street, to work out a better material. We soon got it so that we could make rods in test tubes or boiling tubes and sheets between panes of glass. At this time there was no idea of using a filler and moulding to a final shape."[2]

This last comment is an important one because it confirms that Swinburne and his team did not have sufficient understanding of the nature of the reaction - Baekeland was well ahead - and were unable to exploit the compression moulding technique which was to become a major outlet for the resin.

What they could make, however, was a hard synthetic lacquer for coating brass or other polished metals. With typical humour, Swinburne called this material "Damard Lacquer", a conjoining of the expletive "damn" and the adjective "hard". The word "damn" was regarded by polite society of those days as bad language and, as such, socially unacceptable. It is on record that "a lady secretary" in the company's employ refused to use the trade-name but compromised by using the pronunciation "daymard" and upbraided anyone rash enough to use the proper pronunciation in her presence. By 1909, the total output of lacquer was around 35 gallons a month.

In 1910, there was a significant appointment to the Board of Directors: J.E. Kingsbury, retired Managing Director of the Western Union Telegraph Company. He was to play a major role in the story of Bakelite in the U.K. and is further evidence of the close links that were being forged between the electrical and emerging plastics industries.

It was becoming increasingly clear, however, that the Syndicate was not going to be a commercial success and at an Extraordinary General Meeting convened on 21 March 1910, the company was voluntarily wound up. Its effects and personnel were transferred to a new organisation, The Damard Lacquer Company Limited, with manufacturing premises at 98 Bradford Street, Birmingham.

Fig.4. Artist's impression of the equipment used in the production of phenolic resins at the Bradford Street plant of the Damard Lacquer Company Ltd., c. 1914. Courtesy of the Plastics Historical Society.

The Damard Lacquer Company

It is difficult to believe that the origins of today's vast, international plastics industry stems from companies such as Damard Lacquer Company - but it does.

Damard's early struggles with lack of money, primitive equipment and manufacturing methods, and uncertain sales, mirrors what was going on in other parts of the world. It was to take time, applied science, technology and business acumen to grow the industry into the pattern we see today.

We are particularly fortunate in that one of the great British pioneers in plastics, Howard Vincent Potter, who worked at the Damard Company and was later to become Managing Director and Chairman of Bakelite Limited, placed on record the story of those early struggles. His audio tape recordings made around the late 1950s are a social history document of particular importance to the plastics industry. They are firsthand evidence from a trained chemist. Apart from minor editing and re-arrangement of the order, these are his words:[3]

"I was engaged at an annual salary of £150 by Sir James Swinburne in early 1914 to work in his laboratories at 82 Victoria Street London. My job was to look at three problems: the blooming on Damard Lacquer when used to coat brass (which is still unresolved to this day!); investigate the use of resins other than in the coating field; and examine the feasibility of producing solid product from phenol-formaldehyde resin.

"I succeeded in making a solid material by precipitating the resin after the reaction, drying the sandy-coloured mess and mixing it intimately with a fine wood-flour. The company reluctantly agreed to the investment of five shillings (25p) to make a primitive 'press' - nine inches of gas pipe with a screw plug in one end and a screw-ram device at the other. I filled the pipe with the material, perhaps 4 ounces, screwed down the ram (to compress the mixture) and put it in an oven for about 30 minutes. I then removed it, screwed down the ram again as far as it would go and placed the 'mould' in an oven for another hour. The assembly was then removed, cooled, the screw plug removed and the contents displaced by the screw ram. The product I got was a solid, hard block of material which did not melt, did not show much sign of being a good insulator but did demonstrate that the material could be rendered hard and solid and would not melt again. This was the first moulding (other than by casting) made by the Damard Lacquer Company."

Working Conditions and Personal Problems

Potter was soon to find himself removed to the Bradford Street factory and here his observations concerned the equipment and manufacturing processes. The Bradford Street factory was, in fact, an old house converted for the purpose of manufacture.

"The conditions in Bradford Street were primitive. The laboratory was in an attic at the top and had very little equipment. We made phenol-formaldehyde resin in a shed in the yard which looked on to the entrance to a doctor's house. The fumes from our process led to frequent complaints from both doctor and his patients. Occasionally, the doctor would shout to me 'stop those awful smells'. We would shout back and remind him that the fumes were a good disinfectant - it must be his patients who stank! Eventually, though, I had to have proper ventilation installed to take the fumes into the air above.

"At that time (circa 1915) we were making batches of the resin which we called 'gorm' - a descriptive word for the thick treacle-like resin which on cooling went almost, but not quite, solid. The gorm was cooked in a domestic 20 gallon (90 litre) hot water boiler which had no means of condensation added to it.   The condenser was nothing more than a piece of lead piping inside a water jacket - an inefficient device liable to blow up under pressure drenching everything with hot water, steam and resin!

"The batches of resin were made by filling the tank with the required quantity of phenol - melted out in a tub filled with hot water - poured into the top of the vessel. Formaldehyde was then added, the gas turned on, and the reaction started about 3.00 p.m. It was necessary for me to stand by until the reaction got going and we were satisfied it was under control - at which point the gas would be turned off. The reaction would continue for a time and once the temperature had fallen (oh, yes! we had a thermometer!) the gas was lit again on a low jet and left overnight.   Normally, when you came in the next morning the material had completely reacted and it was run into a large open bath which also had a gas heater underneath.

"The staff to do this work included a girl who stirred the lacquer, another girl to pack it in tins and an old man who transported the tins in a woven basket on wheels and sometimes on public transport. The bad smell of the product got into your clothes, especially woollens, and there is some truth in the story that any worker going home from our little 'tinpot' factory and getting into a tram or bus, would cause all the other passengers to move to the other side!

"Phenol was about 1½ pence per pound - I used to go along to the local gasworks and buy a hundredweight at a time. At that time, of course, 'natural' phenol came from the gasworks as a by-product of coal gas manufacture and its availability depended entirely on the amount of gas being produced. On occasions, I have been along and he said 'Sorry, we haven't got any more - it's all been sold this month. We'll book you for a hundredweight next month.' Formaldehyde was imported, except for a little produced locally, and the war caused the price to rise -which presented us with some difficulty."[4]

Health and Safety?

In another recording Potter revealed what today would be regarded as criminal negligence by the company towards the health and safety of its employees. Chemical fumes apart, he related how a new employee, a Belgian chemist who was a refugee from the war, was entrusted with supervising the phenol-formaldehyde reaction in a vessel located in the basement of the building. The new employee was less than alert and a minor explosion was followed by an escape of the treacle-like resin on to the floor. The unfortunate man's cries for help alerted Potter and another employee who, descending into the basement, saw the chemist stuck ankle-deep in hot resin like a fly in amber. Potter wryly observed:

"the man was very lucky in that he was wearing elastic-sided boots which enabled us to pull him clear leaving the boots stuck fast in the hardening resin. We kept the boots for many years as a treasured memento of the incident."[5]

Another employee, 94-year-old Arthur Lloyd, who was to become Production Director of Bakelite Limited and a distinguished pioneer in his own right, has recorded the fact that the drums of methylated spirit used in the production of the Damard Lacquer was stored under the main wooden staircase of the building - the only route to the floors above.[6] He observed that fires were fairly frequent but all successfully contained by the workforce and the local fire brigade. The "fire-alarm" system comprised a bell attached to the end of a rope to warm people in the upper storeys of the building of impending danger!

The War Years 1914-18

The traumatic years of the First World War were to give rise to three important developments in the history of Bakelite in the U.K. Damard Lacquer had been selling successfully and a decision was made in the autumn of 1912 to start production on Long Island, New York, and sell the product through a company called Universal Castor & Foundry Company. The advent of the 1914-18 war made communications between Britain and the U.S.A. virtually impossible, to which was added a further major problem - Baekeland was suing for infringement of his patents.

It was decided to send Potter to the U.S.A. in 1915 and 1916 to get matters resolved. The decision was made to close the Long Island plant but, more importantly, Potter met and discussed matters with Baekeland. Baekeland, it will be remembered, was a Belgian by birth and he was anxious to assist his homeland in its war with Germany. So he gave Damard permission to use his patents and indicated that, when the war was over, he would come to some business arrangement with them to develop Bakelite in the U.K. Just as important in these deliberations was the fact that Baekeland knew Sir James Swinburne, had seen him work as an expert witness in court, and held him in the very highest regard.

The second development was a spin-off from the war shortages and the inevitable exploitation of science and technology for war aims. Synthetic resins (resinoids) had become essential raw materials - especially for use in laminated insulating boards used in wireless telegraphy on board ships. Damard could not meet the demand for their materials, which far exceeded supply. In 1916, at the invitation of the government's Custodian of Enemy Property, Damard was asked to look over and get into production as soon as possible, a factory at Cowley, Middlesex, which had been run by the German company, Bakelite Geselleschaft, a company in which Baekeland had a controlling interest and among whose European-wide remits had been manufacture and sale of Bakelite resins and products to the U.K. markets.

Anti-German sentiments in World War I were intense and, against this background, the take-over of the Cowley factory was not without its lighter moments. The following is the story in Potter's own words:[7]

"W.R. Cooper, (a fellow director) and myself arrived at the factory
to find a locked gate and an aggressive dog which stopped us from
entering. Eventually, an old man appears, asked us what we

wanted, and said 'you can't come in here ... it's all locked up ... the Germans have left this factory and left the key with me with the understanding that no one is to be let in.'

"When we finally got in, it was obvious that they had first-class equipment, clearly marked as to the function of valves and pipes with everything clean, tidy and polished - as though they thought the war would be over in a few weeks! The German manager (who was a reservist and had returned to his country just before the outbreak of the war) had left 2 cases of personal belongings. These were sealed up by the Admiralty, locked in a cupboard and returned to him some years after the war had finished.

"Our first job was to get a portable steam boiler to link up with the equipment. We found one locally and pulled it into the plant with a horse. Eventually, we found someone to operate the boiler, an old, retired railway engine driver who agreed to take on the job as part of his war effort. We did some test runs in the beautiful copper autoclave" (now preserved in the Birmingham Museum).

"We were not successful at once and made several hard, solid batches of resin which had polymerised (we called it in those days 'a charge of biff') which had to be chipped out before we got anywhere near the product needed by the Admiralty - that is a solid resin that could be dissolved in alcohol and not the liquid form we had been supplying."

The trade name for this resin was FORMITE.

"Incidentally, at the plant, the Germans had been making a resin for casting into umbrella handles, pipe stems and the like. They were cast in lead moulds and the mould-making shop (discontinued during the war) consisted of a container for melting the lead and solid, polished metal dies which were dipped into the molten lead to form the mould into which the resin could be cast. There were a number of these moulds lying about in the shop and the rumour got around that we were making bombs and it ought to be stopped! A local Police Inspector called and was able, after we explained, to satisfy the locals that we were harmless.

"Meanwhile, back at Bradford Street, we were producing no less than three barrels a day. With phenol at 3 3/8 pence [old] per pound and cresylic acid at 2½ pence per pound our product sold at a high price although I can assure you that we never made excessive profits at Damard!"

Fig.5. The production of lead moulds for cast phenolic resin products at the Darley Dale factory. The production ceased around 1928. Courtesy of the Plastics Historical Society.

The Cowley factory was further proof of Baekeland's vastly superior patents and processes. The still referred to by Potter made possible the production of good quality solid resin. Plants also existed for the manufacture of moulding materials by both a wet and dry process. The wet process involved production of a liquid resin which was mixed with fine sawdust or asbestos filler and blended in an industrial mixer. The mixture was then dried and crushed to a coarse powder. In the dry process, solid resin was crushed, ground to a fineness in a ball-mill with fillers and colorants added at the mixing stage. The Damard Company was clearly learning a lot from their wartime activities and this, together with immediate post-war demands for materials, was to lead to the third development: the thought that production should be expanded and concentrated in the Birmingham area. In 1920 a one-acre site was selected in Warwick Road, Greet, and work started on a purpose-built factory to make plastics materials.

The Greet factory could not have come on stream at a worse time. The 1920s slump affected every industry and the company was only able to survive by Sir James Swinburne putting up the necessary collateral from his own resources - a remarkable example of his faith in the product. The new factory enabled Cowley and Bradford Street to be phased out and technical and scientific staff were recruited to work on the problems of producing a moulding material. By 1924 the worst of that recession was over and three major industries - wireless, automobile and electric/telecommunications - were beginning to emerge which called for phenolics as an essential material.

It is at this point, more or less, that Baekeland comes back into the story. Among Damard's competitors in the U.K. were two lively manufacturing companies. Redmanol Ltd. of London were selling agents for excellent phenolic moulding materials and resins made by the Redmanol Chemical Products Company of America. The other was Mouldensite Ltd. of Darley Dale, Derbyshire, who had an exclusive licence to make and sell under the British patent of the Condensite Company of America. Mouldensite's processes were much slower than those of Damard and they limited their business to applications where quality rather than price was of paramount importance. They had one particular significant advantage: they were extremely skilled trade moulders in their own right and could undertake the moulding of complex articles way beyond the know-how of the average trade moulder.

Back in the U.S.A. Dr Baekeland was suing both the American parent companies for infringement of his patents. He was not a vindictive man and his technique, which he successfully carried out, was to acquire a controlling interest and merge them, to the benefit of all concerned, into his own company, the Bakelite Corporation of America, which he registered in the U.K. on 18 March 1926.

Baekeland then approached Damard and, in another display of business acumen, bought a controlling interest in it and, with the agreement of all concerned, merged it with Mouldensite and Redmanol to form Bakelite Limited with factories at Darley Dale and Greet. Sir James Swinburne was appointed Chairman and H.V. Potter became Managing Director. Such was the success that it soon became apparent that the company needed to expand still further and, in 1928, a 10 acre site was purchased at Tyseley, Birmingham, to be followed very shortly by a further 17 acres on the same site - a remarkable testimony to the faith of those early pioneers in their plastics products. Key personnel went to the U.S.A. to study and learn from Baekeland's Bakelite Corporation and the intention, at the outset, was to build the proposed new factory to the highest standards and encourage pride in work, quality of product and job satisfaction. Four more years of planning and construction were to take place before, as the company's historian put it,

> "the first puff of smoke triumphantly emerged from the factory's giant chimney and Bakelite Limited started production."

Like so many other developments of the 1930s, Tyseley and Bakelite Limited are no more, although phenolics are still produced in the world and continue to serve the marketplace. Bakelite generally, however, was overtaken by other polymers offering greater technical advantages and it has to be said for the record that matters were not helped by appalling industrial relations in the 1950s and 1960s.

In 1987, the plant was closed and most of the acres of production buildings fell beneath the bulldozer. So far, the site remains derelict although there are hopes that a new industrial estate will emerge when times become more healthy for business.

To those who worked for the company in its golden years, there will never be a company quite like it. They regarded themselves, and still do, as the Rolls Royce of the plastics industry. Those who are left meet monthly in a Birmingham pub to relive old times and discuss matters such as the collectors craze for Bakelite artifacts. No doubt, too, they will be pleased to hear about this excursion into nostalgia and the good old days when Bakelite was king.

REFERENCES

1. 'Candidate for Election to the Royal Society' (Certificate), Royal Society, London, 29 December, 1905.

2. J. Swinburne, 'Early History of the Damard Lacquer Co.', typed manuscript, undated.

3. H. V. Potter, 'Early History of Bakelite Limited', Reel to reel audio tape (c. 1958).

4. Ibid.

5. Ibid.

6. P. Reboul, 'Interview of A. Lloyd', audio cassette tape, recorded 1 April, 1987.

7. H. V. Potter, 'Early History of Bakelite Limited', Reel to reel audio tape (c. 1958).

Materia Nova: Plastics and Design in the U.S., 1925–1935

Jeffrey L. Meikle

DEPARTMENT OF AMERICAN STUDIES AND ART HISTORY, AMERICAN
STUDIES PROGRAM, 303 GARRISON HALL, UNIVERSITY OF TEXAS AT
AUSTIN, AUSTIN, TEXAS 78712, USA

Colourful new consumer plastics such as cast phenolics, cellulose acetate, and urea formaldehyde appeared on the market just as American commercial design experienced a renaissance stimulated not only by the recession of 1927 but also by an exhilarating mix of European modernisms. The extravagant novelties of Art Deco revealed at Paris in 1925 merged with the austere idealism of Le Corbusier's Towards a New Architecture. Secure in a belief that European artists were celebrating American progress, young designers reacted with equal enthusiasm to the social responsibility of the Bauhaus, the expressionist fantasies of Eric Mendelsohn, the abstractions of de Stijl, and the perceptual machines of Constructivists. More often a matter of gears, motor cars, and "machines for living", modernist visions also encompassed new industrial materials - plastic, aluminium, stainless steel - as of the essence of modernity. Visitors to an avant-garde Machine-Age Exposition in New York in 1927, on viewing Naum Gabo's Constructivist sculptures of intersecting planes of transparent celluloid, realized that plastic, no longer an imitation of ivory or tortoiseshell, lent itself well to machine-age expression.[1]

Among the most vocal boosters of plastic in modern design was Paul T. Frankl, a designer of custom furniture who hoped to see high-art advances extended to the masses. He made a start in that direction in 1928 by designing celluloid comb-and-brush sets described by his client, the Celluloid Corporation, as inspired by "so-called modernistic art".[2] With other emigré designers who had arrived in America since the First World War, Frankl shared a direct knowledge of European modernist trends and an enthusiasm for his adopted country's brash modernity. He was capable of declaring that if the United States had shipped off a skyscraper to the Paris exposition of 1925, "it would have been a more vital contribution in the field of modern art than all the things done in Europe added together".[3] In a book on contemporary design, Form and Re-form, which appeared in 1930, Frankl devoted a chapter to "Materia Nova" that he considered "expressive of our own age". These included such new materials as Bakelite, Celanese, Vitrolite glass, Monel metal, aluminium, linoleum, and cork. Despite the range of this list, he placed greatest emphasis on Bakelite and other synthetics. Warning of conservative clients who sought refuge in imitating traditional materials, he declared that "imagination is essential to visualize and to realize the potentialities of new materials--to treat them on their own terms, to recognize, in a word, the autonomy of new media". He challenged designers "to create the grammar of these new materials" that were already speaking for themselves "in the vernacular of the twentieth century... the language of invention, of synthesis". Frankl announced that "industrial chemistry today rivals alchemy" in applying processes by which "base materials are transmuted into marvels of beauty".[4] Two years later, in another tract on modern design entitled Machine-Made Leisure, he extended his celebration of plastic in a similar chapter whose title announced, "A New Language Emerges". Describing new materials as the foundation of a new style, he again praised the "new alchemy" for yielding the "minute accuracy, strength, permanence, and stability" of moulded plastic. With special praise for William Perkin, whose synthesis of mauve in 1856 had "led the world into fields of loveliness undreamed of before this time", Frankl observed that the future "promises to be a true Plastic Age".[5]

Such visionary rhetoric obscured a tension between the natural and the artificial that remained central to the process of determining forms, textures, and colours of plastic objects during the 1920s and 1930s. A gap divided the pronouncements of someone like Frankl from the actual products available to the ordinary American. While a publicist might declare that Frankl's dresser set of Amerith celluloid "makes one lose faith in the natural, so handsome is the synthetic plastic material", a photograph revealed a surface imitating the rich patterns of onyx. Imitation of the natural or traditional remained a strong motive for using plastic. As late as 1933 an advertisement in Plastic Products proudly declared that "Discriminating Manufacturers Insist on Real Marblette", a cast phenolic illustrated by pictures of three apparently marble ashtrays.[6]

Imitation was the essence of celluloid, of course, but phenolic moulding techniques made it possible to simulate complex three-dimensional forms produced in wood by turners and carvers. The initially drab colour range of moulded phenolic encouraged manufacturers to give plastic objects a bit of warmth by simulating wood. In 1928, before plastic was used for one-piece moulded radios, at a time when decorative wooden cabinets predominated, the editor of Plastic & Molded Products alerted readers to "the possibility of complete molded cabinets, especially those of ornate design, where the cost of carving the wood makes wooden ones expensive".[7] Seven years later, as another observer recalled the ornate "jigsawed" appearance of the first one-piece plastic radio cabinet, moulded by the International Radio Company in 1931, he asked his readers to "imagine a designer's pleasure in knowing that almost any form or amount of ornament he desired could be used safely without the worry of its being loosened or damaged by handling". In that first flush of excitement, he recalled, "no possibility was overlooked". The following year's model had been ornate, "basically gothic in feeling with its pointed arch, grille openings, rosettes and beading". Of the 200,000 sold in 1932, International had moulded three quarters of them in "a mottled effect in imitation of walnut woods".[8]

An executive of General Plastics began speaking out against imitation midway through 1932. Franklin E. Brill was more aggressive than anyone else in promoting plastic as plastic rather than as a simulation of something else. New materials and their uses, he argued, ought to be "frankly expressive of the machine-age which made them possible". Until recently, he and his colleagues had been "mis-using the gifts of our new synthetic age". Without considering plastic's unique possibilities, for example, makers of small electric clocks had released a "flood of imitation walnut Gothic Cathedralettes" so cheap in price and appearance that the market faltered. Seeing bargain counters "loaded with walnut and mahogany moldings" that could hardly be given away had convinced him of the "sheer folly" of "imitating older materials". In his opinion "the consumer subconsciously resent[ed] the manufacturer's low estimation of her intelligence" as revealed in imitations of other materials and simulations of hand craftsmanship. Designers of plastic mouldings would succeed only by treating plastic as a frankly artificial material and by "using simple machine-cut forms to get that verve and dash which is so expressive of contemporary life". Above all, they should abandon ornament - "time-worn motifs" and "fussy, odd shapes - and instead develop plastic's uniqueness by thinking sculpturally in terms of 'contours'". It was necessary "to give our materials an identity" in order to "make them desirable to the consumer and therefore to industry-at-large".[9]

American designers had first become aware of plastic as a modern material when the popularity of radio brought glossy black phenolic laminate into the home in the mid-1920s. They admired this new industrial material for reflecting the machine spirit they hoped to embody in custom furniture for Manhattan sophisticates. Taken together, their work suggested an impersonal precision. Paul Frankl had begun by crafting his unique skyscraper pieces from elegant stained woods, sometimes trimmed with black lacquer, but gradually shifted to Bakelite laminate, at first as an unconsciously ironic imitative substitute for lacquer trim, eventually as a material for surfacing entire pieces. Gilbert Rohde, who started as a designer of custom furniture and ended with commissions for mass-produced goods, produced tables for custom sale with Bauhaus-inspired tubular chrome-steel frames and Bakelite tops. Most active in spreading the use of phenolic laminate was Donald Deskey, whose spare geometric furniture relied for effect on the contrast of brushed metal supports with horizontal expanses of polished black Bakelite so sheer as to embody the abstract concept of two dimensionality. Gaining experience from a room with grey Bakelite walls designed in 1929 for displaying a collection of prints in Abby Aldrich Rockefeller's apartment, Deskey went on to outfit an austere modernist coffee shop for a Brooklyn department store, and in 1932 to specify various laminates on furniture throughout the public rooms of Radio City Music Hall. Phenolic laminate or Formica (as people were beginning to call it generically) quickly spread to all sorts of commercial applications - Woolworth stores, cafeterias, diners, cocktail lounges, Checker cabs, and railway coach interiors. Wherever one looked in the public spaces of the late 1930s, one found the unearthly flawless surfaces of smooth Formica, deep mirrors of jet-black or jade-green or synthetic hues reflecting a futuristic "polished orderly essential" that remained a dream to the real world of the Depression decade.[10]

Achievements of designers like Frankl and Deskey using laminates in custom furniture and interiors alerted executives like Allan Brown, public relations director at the Bakelite Corporation, to design as a factor in promoting plastics among manufacturers reluctant to abandon tradition. Until about 1930 the Bakelite Corporation's promotional strategy revealed a concern for technical over stylistic issues. Sales publications rehashed such familiar points as Bakelite's superiority to natural materials, its uniformity and dimensional stability, lower costs for assembly and finishing, and the company's team of technically trained salesmen. Then in 1930 Bakelite Information, a newsletter for editors and journalists, ran an article on "Bakelite and Beauty in Modern Interior Architecture" with photographs of the splendors of black Bakelite panelling in the Chanin Building, one of the most stylized of Manhattan's new skyscrapers, and of the Formica that graced Brooklyn's even more extravagant Hotel St. George, "from salt water swimming pool to Egyptian roof garden". The article's text praised laminated panelling for the usual technical reasons, then went on to celebrate "the beauty of laminated resinoid... a gloss, a sheen, a depth, and a softness which delight the eye". Although the odd coinage <u>resinoid</u> reeked of the laboratory, Bakelite promotion had clearly entered a new dimension. As employed at the Chanin and the St. George, the material possessed "a beauty not chiefly gained by faithful imitation of wood and marble, but...inherent in the material itself". According to another article in <u>Bakelite Information</u>, now that substitutes had become "better than the original" there was no need "to deceive and often none to produce a resemblance". Plastic could be frankly used as plastic. New materials superior to the old had become a sales point best expressed through a modernist design idiom that promised renewal for a collapsing economy.[11]

Within a short time the Bakelite Corporation sounded more enthusiastic about machine-age design than about Bakelite itself. The company's new strategy got rolling in 1931, a year before American businessmen in general began to discuss design as a Depression panacea. <u>Bakelite Information</u> approvingly quoted Frankl's comments about new materials as a twentieth-century vernacular and plastic as the product of a new alchemy. The newsletter also praised the "modern movement" for offering designers "a new freedom and a release from the historic styles of the past".[12] Bakelite and other plastics were reinvented as up-to-the-minute responses to the spirit of the machine age. Once Brown had adopted a design-oriented perspective from which to promote Bakelite, nothing was too small for notice, not even a phenolic closure for McIlhenny's Tabasco sauce, designed along with its glass bottle by Walter Dorwin Teague and described as "a most piquant little bottle...appropriately capped with a pepper-red resinoid closure". As reported in a Bakelite pamphlet on package design, the new bottle and cap offered a spur to manufacturers "painfully conscious of the need and desirability of more modern design for their containers". But a little bottle, even one whose spicy appearance revealed its contents, hardly encompassed all there was to say about machine-age design and its most expressive material.[13]

Sometime toward the end of 1932, as Americans anticipated inauguration of a new President who promised a new deal, Brown hit on a recovery plan of his own. Not only did it boost Bakelite's reputation as a machine-age material; it also gave tremendous assistance to the fledgling profession of industrial design and awakened designers to plastic's potential as a material of manufacture. On January 5, 1933, Brown convened the first of several seminars intended to acquaint designers with Bakelite's technical and aesthetic advantages. Meeting over luncheon in a private room at a Manhattan hotel, Brown and his staff made an attractive proposition to a dozen of the designers most prominent in restyling commercial products to increase sales by stimulating public interest. Donald Deskey, whose work with laminate had brought design awareness to the Bakelite Corporation, was there. So was Norman Bel Geddes, an extravagant showman whose book <u>Horizons</u>, published two months earlier, was gaining attention both in corporate board rooms and in the Sunday supplements for its vision of a teardrop-shaped utopia.[14] Also attending were Lucian Bernhard, Henry Dreyfuss, Helen Dryden, Lurelle Guild, Gustav Jensen, George Sakier, Joseph Sinel, George

Switzer, John Vassos, and Simon de Vaulchier. Others, among them Raymond
Loewy, Walter Dorwin Teague, and Alfonso Iannelli, soon became involved. The
proposition was simple. If the designers would use Bakelite in actual products
created for their corporate clients, then the company would feature the designers
and their products in an advertising campaign intended to promote industrial design
as a strategy for economic recovery and Bakelite as a mainstay of industrial design.
As an added incentive, the company's engineers would advise designers about
moulding techniques and technical limitations of Bakelite to ensure that the
resulting products proved functionally adequate. Brown later recalled an expansive
goal of "starting the business cycle moving in the opposite direction" and concluded
that the campaign "helped rejuvenate interest in new and better products in
general". As carried out, the plan addressed more specific concerns as well by
publicising industrial design at a formative moment and by establishing Bakelite -
and plastic in general - as a material that both expressed contemporary values and
reduced manufacturing costs.[15]

Fig.6. Raymond Loewy design testimonial for the Bakelite
Corporation, Sales Management, May 15, 1934.

The proposition was irresistible. At least ten designers signed up to appear in advertisements that ran throughout 1933 and 1934. A distinctive format gave the series a strong identity. Each ad focused on a single product and gave its manufacturer free publicity. Each contained a small photograph and capsule biography touting the designer as celebrity, and each quoted the designer on the virtues of sleek modern design, the message being that a product that looks better, sells better. Many of the sketches mentioned other companies a designer had worked for, thereby serving as a source of references for an executive thinking of hiring a particular designer. Those featured ranged from the raffish Ianelli, a goateed artist in shirtsleeves with a cigarette dangling from his lips, to the staid, bankerlike Teague, hat firmly on head, his appearance calculated to reassure the most conservative executive worried about trusting his future to a bohemian artist. An extensive range of products suggested that designing with Bakelite would prove a smart move in any industry. Personal accessories like barometers and telephone indexes, appliances like irons and washing machines, business equipment like mimeographs and soda-fountain dispensers - all were transformed by new materials and new styles. Through it all ran the message, as phrased in one ad, that "distinction in design has a very potent sales appeal". Or, as an ad for a sculptural electric iron phrased it, the product "sheds its Cinderella garb...takes a beauty treatment...and steps out to win housewives and sales".[16]

Appearing in such trade journals as Sales Management and Plastic Products, neither of which boasted an artistic layout, the well-styled advertisements reinforced their own point by contrast. Taken as a whole, the series comprised a primer on product design with an emphasis on plastic. Oddly enough, the texts of the advertisements said little about the actual style that plastic encouraged. Only the photographs suggested how plastic was visually transforming the things of the twentieth century. An alert observer would have discovered a significant trend. Plastic invited the shaping of artifacts as self-contained, irreducible wholes rather than as assemblages of parts. Even moulded-in ornament, most often ribbing or fluting essential for strength, did not appear as something "added on" but instead emphasized the integrity of the single unit. Objects as varied as a telephone index and a meat grinder appeared smooth, rounded, almost sculptural, and made it easy to forget they enclosed mechanical parts. Only one advertisement addressed the issue - and only briefly. It featured a mimeograph machine redesigned by Raymond Loewy and included a small picture of the original for comparison. As exaggerated in description, the old machine was "a mysterious appearing assembly of wheels, gears and cylinders"; the redesigned model, on the other hand, was "a compact machine with operating mechanism concealed within attractive housings".[17] Such a degree of enclosure, reinforced by plastic's real and visual seamlessness, marked a break with much of civilization's prior material and mechanical evolution. Until the advent of plastic, most objects appeared as assemblages of various parts and materials; they did not seem smoothly, inviolably whole. Wood, for example, invited makeshift repairs, modifications, additions. Iron and steel proved less open to manipulation by ordinary people, but even a suspension bridge or the Eiffel tower could be understood as a complex assemblage of relatively simple parts. Plastic, on the other hand, fostered a "black box" syndrome of ignorance about technological processes by enclosing them in irreducible moulded forms whose deceptive simplicity found its earliest, clearest expression in the streamlining of the 1930s. It became easy to look no further than a flawless surface, whether geometrically precise or sensuously flowing, and to assume that beneath it lay an ideal state of perfection. Ironically, artifacts shaped by plastic during a self-conscious machine age indicated a willingness to ignore mechanical complexities, to abdicate responsibility for understanding or directing them, to assume that beneath those pristine surfaces everything was well under control. Plastic contributed in the most fundamental way to a victory of style over substance in the minds of consumers whose millions of individual decisions conjured up artificial environments to which they then had to adapt. Style, it turned out, could be quite substantial.

The campaign to link Bakelite's fortunes to industrial design succeeded. Free publicity and direct experience with new materials inclined designers to use plastic in the future. Manufacturers who employed designers proved receptive to their suggestions that plastic be adopted because the advertisements had shown what it could do. At times the strategy backfired, as when manufacturers of two of the campaign's star products abandoned Bakelite and began moulding the same products from Durez phenolic. Unlike most American industries during the Depression, however, the plastic industry accommodated competition by continuing to expand. Whether used to reduce costs or to appeal to a public seeking intimations of future progress, the combination of plastic and design proved irresistible. The Bakelite Corporation continued to promote plastic as a consumer material by publishing a portfolio on design and designers, possibly a compilation of the series of advertisements.[18] The Bakelite Review began printing names and addresses of approved designers who had demonstrated adequate knowledge of the material's technical and aesthetic potential; in 1939 the company consolidated these names in a list distributed on demand to manufacturers.[19] Annual sales of Bakelite moulding compounds in the meantime rose from 8.3 million pounds in 1931 to 30.4 million in 1937, and that division of the company emerged from red to black in 1935. If the design campaign was not entirely responsible, it had certainly contributed.[20]

Other plastic companies quickly followed Bakelite's lead in embracing modern styling. Some hired consultant industrial designers to work with specific clients who wanted to use their materials and in general to develop new applications. Gilbert Rohde, for example, consulted with Rohm & Haas during its introduction of Plexiglas. Even custom moulders sometimes employed designers. The Kurz-Kasch Company of Dayton, Ohio, maintained a long association with the design partnership of Carl Sundberg and Montgomery Ferar, who provided prototypes for convincing reluctant customers to switch to plastic. James Barnes and Jean Reinecke enjoyed a similar relationship with Chicago Molded Products beginning in 1935. Reinecke later attributed their success to the fact that they knew a bit more than any of their customers about an esoteric subject. "The fact that we could glibly say "methyl methacrylate", as well as "Bakelite", was very authoritative", he recalled, leading not just to success with Chicago Molded Products but also "to our writing various 'how to' articles" for trade journals that in turn generated further design commissions in plastic.[21] Nothing better signified the American plastic industry's shift from technical to aesthetic matters than the transformation late in 1934 of its trade journal, restyled and renamed Modern Plastics with a new editorial policy that gave as much space to designers as to plastics executives or engineers. By then the journal hardly seemed innovative; instead it reflected an attempt to catch up with Bakelite's lead and reap the benefits for the entire industry.

Plastic's expansion into consumer goods occurred just as designers and architects were developing streamlining as a machine-age design mode more expressive of Depression aspirations than the rhythmic angularity of 1920s Art Deco. As Peter Muller-Munk argued, with the appearance of plastic's "large flowing curves", "the purely façade type of design" was "giving way to a really plastic conception of the machine as an object to be seen from all angles at last".[22] Low, horizontal, sculptural, flowing, evocative of speed, streamlined design reflected desire for frictionless flight into a utopian future whose rounded vehicles, machines, and architecture would provide a visually uncomplicated, protective environment - closed off from the Depression's economic and social dislocations and marked by a static perfection. A form of cultural expression generated by aerodynamic research and by the prominence of the automobile, streamlining also enabled the most economical development of plastic moulding techniques. The most typical moulded consumer product was the plastic radio cabinet, which physically embodied a revolution in communication similar in scope to that effected by the automobile in transportation. While ordinary people might look with longing at Airflow Chryslers and Lincoln Zephyrs, they could actually afford a Sears Silvertone, a Pilot All-Wave, or a two-tone Fada. The plastic radio was the most democratic exemplar of streamlining, a machine-age icon for the living room of everyman and everywoman. And moulded phenolic, the decade's most typical plastic material, evoked utopian permanence with its hard smoothness, its flawless surface, its quality of being "anything but plastic, once it has taken shape".[23] However accidental the convergence of the particular material plastic with the larger cultural expression of streamlining, the relationship became as symbiotic as that of the plastic industry with industrial design.

The industry's transformation over two decades owed success to a convergence of new materials and new consumer-oriented marketing techniques such as industrial design. Promoters like Allan Brown of Bakelite had worked hard to educate manufacturers, and then consumers, to perceive plastic's worth. Once the message got through, some manufacturers began demanding more than plastic could deliver. Sometimes requirements of a particular application stimulated development of a new material. What might be called "designer materials" based on polymers that at least in theory were built to order began to replace exploitation of fortuitous discoveries.[24] The story of Plaskon, the urea formaldehyde moulding compound announced in 1931, not only suggests this responsiveness of plastic to outside cultural demands but also summarises the complex interaction of scientific research, technological innovation, and marketing strategy that typified the industry during these years of emergence.[25]

Plaskon owed its development to specific needs of the Toledo Scale Company, which had been making commercial weighing scales at Toledo, Ohio, since 1901. After twenty-five years under the leadership of its founder, the company gained a new president in 1926. Hubert D. Bennett, a 1917 graduate of Williams College, had worked in advertising and as director of sales for Studebaker in Brooklyn before arriving in Toledo. Among the problems confronting Bennett was a cast-iron grocery counter scale whose weight prevented all but the burliest salesmen from carrying it. Grocers who bought it on the basis of sales literature often complained that its weight kept them from rearranging their stores--a prime strategy of machine-age retailing. Before 1923 the scale had weighed in at seventy pounds and was a popular seller. In that year the company had switched from lacquer to porcelain enamel to obtain a more durable finish. Because porcelain had to be fused to iron at 750'C, the iron parts had to be thickened and ribbed to prevent their warping under the heat, and the scale's weight had jumped to 163 pounds. Bennett, willing to take risks to rejuvenate the company, attacked this problem of marketing and consumer demand by simultaneously seeking a new metal and a new coating. As an associate observed at the time, he was "interested in perfecting entirely new ideas, so he can have them patented".[26]

Fig.7. Cast-iron grocery counter scale for which the Toledo Scale Company sought a replacement. Photograph courtesy of Harper Landell.

Bennett's first move came in December 1928, when he engaged designer Norman Bel Geddes to work on several scales, including the cylinder scale for grocers. The two had met over a tennis net in Toledo when Geddes was visiting a cousin who had known Bennett at Williams College; Mrs. Stanley Resor, wife of the president of the J. Walter Thompson advertising agency, brought them together again in the hope that Bennett would employ Geddes. They got on so well that within six weeks of signing the initial contract Bennett was asking Geddes to design a new factory in which to make the scale. Work progressed on both fronts, with sheet steel the material of choice for the scale. Geddes's draftsmen produced a series of striking charcoal renderings reminiscent of grain elevators and other machine-age forms - one of which brought considerable publicity when it appeared in Fortune.[27] In the meantime Bennett initiated an investigation at the Mellon Institute of Pittsburgh into the possibility of developing a lightweight durable coating that could be used on metal scales in place of porcelain enamel. When preliminary research indicated water-white urea formaldehyde as a potential resin for such a coating, Bennett funded an industrial fellowship to investigate it systematically. Chemist Arthur M. Howald formally set to work on the problem in January 1929.[28] Although Bennett intended the two efforts to converge in a new scale, the situation became more complex.

Each of the lines of inquiry Bennett set in motion began to lead in its own way to the radical notion of abandoning metal entirely in favour of using plastic for the new scale's housing. As Geddes told the story some years later, he became fascinated by a round glass window through which shoppers could watch the operating mechanism of the old cast-iron scale. Why not make the whole thing transparent, he thought, and if glass was too fragile, then one of those new plastics might work. It is a sign of the mingled ignorance and wonder surrounding plastic that Geddes was surprised to learn that a large transparent Bakelite moulding was out of the question.[29] His enthusiasm for plastic ironically cost him his job. Bennett found the idea of a plastic scale so fascinating that he abandoned Geddes's sheet-metal plans to focus full attention on new materials even if it meant a delay in replacing the old cast-iron millstone. When he discovered Bakelite was the only plastic strong enough for his purposes, and that it could be had only in black or dark brown, he had ten black scales installed in groceries around the country to test shoppers' reactions to weighing food on black. Not only did they prefer the "sanitary appearance of the white machines", the study revealed; many of them "refused to buy food weighed over the black scales".[30] This rejection might have sent him back to Geddes's sheet-metal designs. But then Howald at the Mellon Institute suggested using urea formaldehyde not as a resinous coating but as a plastic moulding compound indistinguishable from Bakelite except in its crucial capacity for accepting any desired colour - including white.

After solving problems of water absorption and fading in sunlight, Howald's work yielded Plaskon, a cellulose-filled urea formaldehyde moulding compound produced in hundred-pound lots in a pilot plant at Mellon in 1930. Recognising potential uses beyond scale housings, Bennett formed the Toledo Synthetic Products Company under the presidency of James L. Rodgers Jr., another 1917 Williams graduate. Commercial production began in April 1931 in a plant at Toledo under Howald's supervision. From the start they promoted Plaskon as a material especially suited for consumer goods. Rodgers observed that Plaskon, removing any need "for cheap imitation of woods and marbles", would initiate a new type of design in plastic "relying rather on the intrinsic beauty of the material itself". An arty promotional booklet of 1934, illustrated with luminous tinted photographs of Plaskon objects printed on black, was called simply Plaskon Molded Color.[31] By then Bennett had hired a new design consultant, Harold Van Doren, both to promote Plaskon with visually effective prototypes and to design the housing for a new grocery counter scale of moulded Plaskon. Another graduate of Bennett's class at Williams, Van Doren had studied at the Art Students League and in Paris, and had worked as an editor before taking a position in 1927 as assistant director of the Minneapolis Institute of Art, where he remained until 1930 when he set up shop as a designer at Toledo. It seems likely that classmates Bennett and Rodgers talked him into making the move. If so, then one of the most articulate of the new industrial designers owed his career to plastic.[32]

Early Plaskon mouldings designed by Van Doren and his partner John Gordon Rideout introduced the material to manufacturers. Among them were lids for cosmetic jars, closures for tubes, small clock cases, a salt shaker resembling a Buck Rogers observatory, and a small biscuit cutter given away by the millions in boxes of Bisquick. Designing a one-piece moulded case for the Air-King radio of 1933 gave Van Doren his first experience with a large housing. A foot high, nine inches wide, and nearly eight inches deep, the moulding taxed manufacturing processes but ultimately proved technically feasible and visually attractive to consumers. Popular response to this modernistic radio perhaps suggested to Van Doren and Bennett that a Plaskon scale should "look as different from the scale made of cast iron as the modern automobile does from the buggy it replaced".[33] The old cast-iron scale was an ungainly, top-heavy profusion of visually complex parts. By contrast Van Doren's new design was simplicity itself. According to his presentation rendering, a low rectangular case enclosed both the works and the revolving cylinder from which weights were read; the weighing platform discretely hugged the top. Receding bevels rounded the upper half of the moulding, provided visual interest, made it easier to read the cylinder from a high angle, and strengthened the moulding. About as wide as the old scale at eighteen inches, Van Doren's new design was less than half as deep and less than half as high at about a foot. Its smooth horizontal enclosure of the mechanism within a single volume embodied the public's passion for streamlining. But there was a big problem. Only after the design was completed, and the mechanism substantially re-engineered as well, did Bennett discover that no one could mould such a large piece, that in fact, if successful, the scale housing would be "the largest plastic molding ever made in the United States".[34]

By then, having funded development of a new plastic to reach the goal of a lighter, less bulky scale, Bennett refused to give up. He entrusted manufacture of the scale housing to General Electric's custom moulding department. The job required constructing the largest hydraulic moulding press then in existence, a task carried out by the French Oil Mill Machinery Company of Piqua, Ohio. When installed in GE's plant at Fort Wayne, Indiana, the new press stood twenty-two feet high, weighed eighty-nine thousand pounds, and exerted fifteen tons of pressure. The mould was cut from a seven-ton block of steel. These oft-repeated statistics suggested the awe with which mechanical engineers regarded the accomplishment; a drawing of the press used in technical journals depicted it from a point near ground level as a behemoth towering over a gnomish figure in laboratory garb.[35] Despite this image of mastery, the moulding process proved difficult. Some pieces warped when removed from the mould. Many revealed cosmetically unacceptable flow lines. Others lost their polished surfaces when pulled away from the mould. Dust, not a factor in smaller mouldings, contaminated the smooth, white surfaces. The press room had to be air conditioned and elevated pressure maintained to keep air flowing out rather than in. The very size of the equipment brought unexpected delays when it had to be repaired. Despite all these problems, the momentum established by Bennett prevailed. In August 1935 Toledo Scale began taking orders for its Duplex model, later known as the Sentinel.

The new scale more than met Bennett's initial goal. With a total weight of fifty-five pounds (the case itself weighing about eight pounds), the Sentinel could be moved easily by salesmen and grocers. Its smooth lines radiated efficiency. Sales increased by three hundred percent in six months as grocers discarded cast-iron reminders of the Victorian era. Success helped offset development costs of some $500,000 and the use of more expensive materials. While cast iron cost only four cents a pound, Plaskon cost thirty-five cents, and the aluminum used for most working parts cost forty cents; Toledo Scale thus paid three times as much for the materials in a Sentinel. Offsetting that increase were considerable reductions in assembly and shipping costs. Less than a third as heavy as the old scale, the Sentinel incurred lower freight charges and was packed in a cardboard box rather than a wooden crate. Beyond such corporate concerns, the success of Bennett's experiment convinced other manufacturers to abandon traditional materials for moulded plastic. He exaggerated when he claimed the new scale as "the first time a synthetic chemical product [was] brought into real competition with the old established mined or refined substances".[36] But the Sentinel did give other manufacturers and designers courage to adopt similar large mouldings of urea or phenolic for an array of housings, cases, and cabinets. Soon after its introduction, an engineer at General Electric predicted that "business machines, radios, electric apparatus, vending machines, recording instruments", and food-handling equipment would soon sport plastic housings.[37] Toledo Scale's experience made it possible to take plastic seriously as a material.

Fig.8. Duplex or Sentinel grocery scale with a Plaskon housing introduced by the Toledo Scale Company in 1935. Photograph courtesy of Harper Landell.

Bennett and his associates recognised that more than size distinguished their accomplishment from earlier uses of plastic. Faced with a specific need for a new material light in weight and colour, they had followed a complex trail of development and innovation without regard for expense or traditional wisdom. In the process they had created not only a new material, Plaskon, but also a method for moulding it and a visual style appropriate for its presentation to consumers. Applied chemical research, production engineering, and the marketing strategy of industrial design came together to offer a solution that none alone could have provided. No longer a novelty, plastic was emerging during the 1930s as a staple essential to other industries. No one was ready to take it for granted, however. Ordinary consumers, who lacked a sense of plastic's economic justification and who experienced it as moulded by the streamlining medium of industrial design, had no reason to question its utopian force. As for businessmen, the many journal articles devoted to Plaskon and the Sentinel scale testified to their continuing, even intensified regard for plastic as something of a miracle substance. The quick shifts from cast iron to sheet metal to plastic offered a paradigm for other industries and products. Only later, however, did the extent of plastic's subversive effect on the design of the material landscape--its potential for malleability--become fully apparent. The official utopia of the late 1930s was reflected in the buildings of the New York World's Fair of 1939, whose rounded streamlined forms actually promised static closure. Moulded radios, clocks, telephones, and grocery scales echoed the official vision of stability by their styling. Their durable material substance also promised stability, even permanence, especially when compared with things subject to rot or corrosion. But the process by which plastic was proliferating in the material environment promised anything but stability--a fact that became apparent only with the material abundance of the 1950s and 1960s and the new thermoplastic materials that embodied it.

REFERENCES

1.   On the Machine-Age Exposition in general see Dickran Tashjian, 'Engineering a New Art', in 'The Machine Age in America 1918-1941' by R. G. Wilson, D. H. Pilgrim, and D. Tashjian , Harry N. Abrams, New York, 1986, pp. 231-235.

2.   'Modern Art in Pyroxylin', Plastics & Molded Products, 1928, 4, December, 695.

3.   P. T. Frankl, 'New Dimensions: The Decorative Arts of Today in Words and Pictures', Payson and Clarke, New York, 1928, p. 61.

4.   P. T. Frankl, 'Form and Re-form: A Practical Handbook of Modern Interiors', Harper, New York, 1930, pp. 31, 163.

5.   P. T. Frankl, 'Machine-Made Leisure', Harper, New York, 1932, pp. 112, 115, 117, 122, 120.

6.   'Christmas Gifts for All', Plastics & Molded Products, 1930, December, 696; advertisement for the Marblette Corp., Plastic Products, 1933, 9, October, 305.

7.   C. Marx, 'How Changes in Style and Design of Radio Apparatus May Affect the Plastics Industries', Plastics & Molded Products, 1928, 4, October, 566.

8.   G. Voss, 'Practical Plastics Score Over Traditional Wood', Modern Plastics, 1935, 12, March, 37, 68.

9.   Quotations are drawn from F. E. Brill, 'Our Homesick Plastics', Plastics & Molded Products, 1932, June, 235-236; Brill, 'Some Hints on Molded Design', Plastic Products, 1933, 9, April, 54-55; and Brill, 'Phenolics for '36 Clocks', Modern Plastics, 1935, 13, November, 27. See also Brill, 'Midget Radios Versus Electric Clocks', Plastic Products, 1933, 9, July, 182.

10. The quoted phrase is a journalist's paraphrase of Raymond Loewy in 'Streamlining - It's Changing the Look of Everything', Creative Design, 1935, 1 , Spring, 22. On designers' use of phenolic laminate see W. R. Storey, 'Plastics Enter the Home', House Beautiful, 1933, 74 , December, 276-278, 291-292; E. F. Lougee, 'Furniture in the Modern Manner', Modern Plastics, 1934, 12, December, 18-20, 61-62; E. F. Lougee, 'Planning Ahead with Gilbert Rohde', Modern Plastics, 1935, 12, July, 14; Diane H. Pilgrim, 'Design for the Machine', in 'The Machine Age in America 1918-1941', especially pp. 276-303; and David A. Hanks with Jennifer Toher, 'Donald Deskey: Decorative Designs and Interiors', E. P. Dutton, New York, 1987.

11. 'Bakelite and Beauty in Modern Interior Architecture', Bakelite Information, 1930, 14, June; Fred C. Bowman, 'Replacement versus Substitution', Bakelite Information, 1930, April.

12. 'Resinoid, Metal, and Wood', Bakelite Information, 1931, 17, March; 'Modern Art in Industry', Bakelite Information, 1931, 23, December.

13. Zaida A. Ellis, 'The March of the Molded Closures', Plastics & Molded Products, 1930, 6, September, 527; 'Restyling the Container to Increase Sales', Bakelite Corporation, New York, 1931: 16 pp., J. Harry DuBois Collection, box 1, Archive Center, National Museum of American History, Smithsonian Institution, Washington, DC.

14. On the impact of Geddes, 'Horizons', 1932, Little, Brown, Boston, see J. Meikle, 'Twentieth Century Limited: Industrial Design in America, 1925-1939', Temple University Press, Philadelphia, 1979, pp. 144-148.

15. Allan Brown interview, 1952, Plastics Pioneers Association Tapes, reel 1, side 1, Archive Center, National Museum of American History, Smithsonian Institution, Washington, DC (hereafter 'PPA Tapes'); Allan Brown interview, December 26, 1974, tapes provided by William T. Cruse. See also 'Industry Is Having Its Face Lifted', Bakelite Information, 1933, 31, February; 'Plastics in Pictures', Plastic Products, 1933, 9, March, between pp. 22-23; and J. Harry DuBois, 'Plastics History U.S.A.' Cahners, Boston, 1972, pp. 184-185.

16. Advertisements quoted are from Sales Management, 1933, August 15, 171; 1934, March 15, 237. Ellipses in original.

17. Sales Management, 1934, May 15, 485.

18. 'Letters: Our Portfolio on Product Design', Bakelite Review, 1934, 6, October, 1.

19. See Bakelite Review throughout 1938; 'Designers', Bakelite Review, 1939, 11, October, 14.

20.   L. C. Byck, 'A Survey of the Bakelite Thermosetting Business 1910 Through 1951', 1952, November 3, p. 22, <u>Leo H. Baekeland Collection</u>, division VI, box 12, folder VI-B-6, Archive Center, National Museum of American History, Smithsonian Institution, Washington, DC. Byck did not report figures for 1932; they were almost certainly lower than those for 1931.

21.   'News', <u>Modern Plastics</u>, 1939, <u>16</u>, February, 56; Robert L. Davidson of Kurz-Kasch interviewed by the author, 1985, July 22; and letter from Jean Reinecke to the author, 1985, June 7.

22.   P. Muller-Munk, 'Vending Machine Glamour', <u>Modern Plastics</u>, 1940, <u>17</u>, February, 66.

23.   'What Man Has Joined Together...', <u>Fortune</u>, 1936, <u>13</u>, March, 71.

24.   One of the first to suggest this was Archie J. Weith of Bakelite, as quoted in 'A Plastic a Day Keeps Depression Away', <u>Chemical & Metallurgical Engineering</u>, 1933, <u>40</u>, May, 248.

25.   Discussion of Plaskon and the Toledo Scale Co. is based on A. M. Howald, 'Systematic Study Develops New Resin Molding Compound', <u>Chemical & Metallurgical Engineering</u>, 1931, <u>38</u>, October, 583-584; J. L. Rodgers, 'Plaskon, a New Molding Compound the Result of Planned Research', <u>Plastics & Molded Products</u>, 1931, <u>7</u>, December, 664-665, 687; N. Bel Geddes, 'Horizons', Little, Brown, New York, 1932; 'Redesigned Scale, with Molded Plastic Housing, Is Tribute to Research', <u>Steel</u>, 1935, <u>97</u>, August 5, 75; 'Research - on a New Scale', <u>Business Week</u>, 1935, August 10, pp. 9-10; 'Giant Plastic Molding Press Produces Large Weighing Scale Housings', <u>The Iron Age</u>, 136 (August 29, 1935), 13-14; H. D. Bennett, 'Pushing Back Frontiers', <u>Modern Plastics</u>, 1935, <u>13</u>, September, 25-27, 30-32; S. A. Maxom, 'Hail to the Scale', <u>The Du Pont Magazine</u>, 1935, <u>29</u>, October, 8-9; N. S. Stoddard, 'Molding the Toledo Scale Housing', <u>Modern Plastics</u>, 1935, <u>13</u>, October, 11-13, 56-57; 'Plaskon: A New Plastic Material', <u>Machinery</u>, 1935, <u>42</u>, November, 169-174; E. R. Weidlein and W. A. Hamor, 'Glances at Industrial Research During Walks and Talks in Mellon Institute', Reinhold, New York, 1936, pp. 59-82; 'What Man Has Joined Together...', <u>Fortune</u>, 1936, <u>13</u>, March, pp. 74-75; 'Modernized Chopper', <u>Business Week</u>, 1936, August 8, 16; H. Van Doren, 'Industrial Design: A Practical Guide', McGraw-Hill, New York, 1940, pp. 48, 171-172, plates 16-18; W. Haynes, 'American Chemical Industry: The Chemical Companies', D. Van Nostrand, New York, 246-247; 'Hubert Bennett Stressed Research in Industry', <u>Toledo Blade</u>, 1951, September 9, clipping, Toledo-Lucas County Public Library; 'Hubert D. Bennett: Industrialist Once Headed Toledo Scale', <u>Toledo Blade</u>, 1951, September 9, clipping, Toledo-Lucas County Public Library; 'J. L.Rodgers Jr., Plastic Leader', <u>New York Times</u>, 1955, February 26, p. 15; 'H. Van Doren: Industrial Designer Was Ex-Toledoan', <u>Toledo Blade</u>, 1957, February 4, clipping provided by Harper Landell; J. Harry DuBois interview, 1968, PPA Tapes, reel 2, side 2; and DuBois, 'Plastics History U.S.A.', Cahners, Boston, 1972, pp. 117-121, 159-160, 190-191. I also relied on manuscript materials in the <u>Norman Bel Geddes Collection</u>, Hoblitzelle Theatre Arts Library, Harry Ransom Humanities Research Center, University of Texas at Austin (hereafter 'Geddes Collection'); on photographs and documents

in the possession of Peter Bressler, Philadelphia, and Harper Landell, Downingtown, Pennsylvania; and on an interview with Bud Kasch, July 22, 1985.

26. Typed job summary for Toledo Scale Co., <u>Geddes Collection</u>, file 152.

27. 'Bel Geddes', <u>Fortune</u>, 1930, <u>2</u> , July, 51, 54, 57. Other information is from the contract between Bennett and Geddes, December 18, 1928; 'Record Copy Book' for Toledo Scale Co., 1928; typed job summary for Toledo Scale Co.; all <u>Geddes Collection</u>, file 152; and Geddes, 'The Horse Race Game', transcript, 1951, December 2, <u>Geddes Collection</u>, Autobiography source file, Chapter 47.

28. Haynes gave the date of the fellowship's inception as January 1, 1928. On p. 64 of their in-house history of the Mellon Institute, Weidlein and Hamor gave it as January 1, 1929; written with Bennett's assistance at an earlier date, their account is likely more accurate. Queries addressed to several librarians and archivists at Carnegie Mellon University failed to uncover records of the industrial fellowships.

29. Typed transcript of telephone conversation of Geddes with a Mr. 'Edgerter', an executive in the plastic industry, 1954, August 27, Geddes Collection, Autobiography source file, Chapter 40.

30. E. R. Weidlein & W. A. Hamor, 'Glances at Industrial Research During Walks and Talks in Mellon Institute', Reinhold, New York, 1936, p. 63.

31. J. L. Rodgers, 'Plaskon, a New Molding Compound the Result of Planned Research', <u>Plastics & Molded Products</u>, 1931, <u>7</u>, December, p. 687; <u>Plaskon Molded Color</u> (Toledo: Toledo Synthetic Products Co., 1934), 20 pp.

32. Biographical information is from a one-page curriculum vitae in the possession of Harper Landell, who took over Van Doren's practice when he died in 1957.

33. E. R. Weidlein & W. A. Hamor, 'Glances at Industrial Research During Walks and Talks in Mellon Institute', Reinhold, New York, 1936, p. 68.

34. Ibid., p.78.

35. Reprinted in DuBois, 'Plastics History U.S.A.', Cahners, Boston, 1972, p. 119. Compare with a photograph on p. 160.

36. E. R. Weidlein & W. A. Hamor, 'Glances at Industrial Research During Walks and Talks in Mellon Institute', Reinhold, New York, 1936, p. 81.

37. N. S. Stoddard, 'Molding the Toledo Scale Casing', <u>Modern Plastics</u>, 1935, <u>13</u>, October p. 56.

Synthetic Rubber: Autarky and War

Peter J. T. Morris

COLLECTIONS DIVISION, THE SCIENCE MUSEUM, EXHIBITION ROAD,
LONDON SW7 2DD

We are living in the "Polymer Age", and few polymers have been more important than synthetic rubber. World production reached 900,000 tonnes as early as 1944; it is now just over ten million tonnes a year, about twice the output of natural rubber. This scale of production made synthetic rubber a pioneer in the field of high-tonnage polymers. Synthetic rubber research stimulated the development of other polymers, notably polystyrene, acrylonitrile-butadiene-styrene (ABS) and acrylic fibres.

THE ORIGINS OF BUNA S

At the beginning of the twentieth century there was a rising demand for rubber, in particular for pneumatic tyres in bicycles and the new-fangled motor-car. The supplies of wild rubber from the Amazon basin were clearly inadequate and plantations sprang up in the Far East to meet the demand. Rubber trees take several years to mature, however, and the rubber price continued to spiral upwards. The all-time high of 12s 9d (64p) per pound was reached in 1910. The plantations now came into production and the price slumped to less than 3s (15p) per pound in 1913. In the meantime these high prices stimulated synthetic rubber research in several countries.

Superficially Germany was in an excellent position to develop synthetic rubber. It lacked a secure supply of natural rubber, but possessed abundant reserves of coal for organic synthesis. Between 1863 and 1914, the German dye industry had created formidable industrial research and development systems based on German excellence in organic chemistry. By 1914, the German dye firms had successfully developed synthetic analogues of the natural dyes alizarin and indigo, countless synthetic colours, and a growing number of therapeutic agents, including tuberculosis antitoxin, barbiturates, aspirin and Salvarsan. They had also built up good sales organisations for dyes and pharmaceuticals. Furthermore German academic chemists were at the forefront of the new fields of physical and colloid chemistry and were in a strong position to develop the scientific study of polymers.

Bayer was the first German company to tackle the synthesis of rubber. The impetus came from a surprising source, a speech made by Wyndham Dunstan of the Imperial Institute to the Chemical Section of the British Association at its meeting at York in 1906, entitled 'Some Imperial Aspects of Applied Chemistry'.

In the course of his discussion of rubber chemistry, Professor Dunstan remarked, "It cannot be doubted that chemical science will sooner or later be able to take a definite step towards the production of rubber by artificial means". Fritz Hofmann, chief chemist in Bayer's Pharmaceutical Department, read a report of Dunstan's speech in *Chemiker-Zeitung*, and immediately asked Carl Duisberg for funds to pursue this line of research. Duisberg provided a grant of 100,000 marks a year, and a small research group was set up.

Hofmann initially concentrated on the synthesis of isoprene. He successfully produced polyisoprene in 1909, but his process was quite uneconomic, owing to the high cost of synthetic isoprene. Not long afterwards, Hofmann's assistants, Kurt Meisenburg and Curt Delbrück, prepared methylisoprene by dimerising acetone with aluminium. Acetone was a relatively cheap product obtained from wood distillers, and it was manufactured from acetylene by Hoechst in 1916. Bayer adopted the seed polymerization noted by Ivan Kondakov. A small piece of polymer was added to a tank of methylisoprene, which was converted in a period of several weeks into a cauliflower-like rubber. This methyl rubber was first used (mixed with natural rubber) to make tyres in 1912, and was used as an *ersatz* material in the First World War. Bayer made over 2,000 tonnes of methyl rubber during the war, and BASF a much smaller amount using sodium polymerisation. It was leathery, aged rapidly and was unsuitable as a tyre rubber; vehicles with methyl rubber tyres had to be jacked up overnight. As it was also three times more expensive than natural rubber, it is not surprising that the production of methyl rubber was abandoned at the war's end.

If the falling price of natural rubber, which declined by seventy-six per cent between February 1920 and March 1922, made synthetic rubber uneconomic, it also threatened to bankrupt the rubber plantations in British Malaya. In 1922, Winston Churchill, then Colonial Secretary, introduced a plan (the so-called Stevenson scheme) to restrict rubber exports from Malaya to force up the price and thereby protect the infant industry. This produced predictable upheavals in the market and the rubber price suddenly accelerated to 4s 5d (22p) per pound in July 1925. To calm the situation, the British authorities raised the exportable percentage in November, but the damage was done. The major rubber consuming nations, enraged that they were at the mercy of the British government, encouraged research into synthetic rubber. The period between 1925 and 1932 saw the establishment of modest synthetic rubber production in the Soviet Union, Germany and the USA.

In 1925, the major German dyestuff companies, including BASF, Bayer and Hoechst, merged to form IG Farbenindustrie AG (usually shortened to IG Farben or IG). The new concern decided, as a result of the rubber panic and with the encouragement of Germany's largest rubber company, Continental Gummiwerke AG (Conti), to resume research into synthetic rubber. Initially attention was focused on the sodium polymerisation of butadiene to produce polybutadiene or "Buna" (a contraction of Butadien and Natrium, the German names for butadiene and sodium). While Buna rubber represented a significant improvement on methyl rubber, its low tear and tensile strength, and its stickiness, made it unsuitable as a passable tyre rubber.

The former Bayer chemists in Leverkusen strove to replace sodium polymerisation with emulsion polymerisation. In emulsion polymerisation, the hydrocarbon starting materials are suspended in an aqueous emulsion by soap, and the resulting polymer is in the form of small particles or "crumbs" rather than a solid block. This allows a greater degree of control over the operating temperature and the final product is easier to handle. Buna S arose from an attempt by Claus Heuck in Ludwigshafen to "lubricate" the unsatisfactory "dry" and crumbly polymer produced by the emulsion polymerisation of butadiene by adding drying oils, usually linseed oil, before the polymerisation. Kurt Meisenburg, who was working on the polymerisation of styrene in Leverkusen, then discovered that a small amount of styrene could produce a similar improvement. In June 1929, his colleague Walter Bock polymerised butadiene and styrene in a 2:1 ratio, to produce a copolymer with excellent wear resistance when compounded with carbon black. The key patent was filed on 21 July 1929.

By a remarkable coincidence, Hermann Mark and Carl Wulff, in Ludwigshafen, patented a cheap route to styrene 19 days later. Now Buna S (S for Styrol, the German name for styrene) was not only better, but also cheaper than the sodium-based rubbers. Buna N, a copolymer of butadiene and the more expensive acrylonitrile (the N stands for Nitril), was patented by Leverkusen's Erich Konrad in April 1930. By this time, IG was concerned about the sharp fall in the price of natural rubber and the economic effects of the Depression. Six months later, before Buna S and Buna N had left the laboratory, the synthetic rubber research was practically suspended.

Fortunately tyre tests at the Nürburgring racecourse were permitted to continue and they revealed in December 1931 that Buna S was significantly more hard-wearing than natural rubber. It also became clear that it was impossible to process Buna S or Buna N on the machinery used by the rubber industry, because they were harder than natural rubber. In February 1932, IG decided to find an American rubber company to collaborate on experiments to overcome this problem, through its American ally Standard Oil of New Jersey. Negotiations with Goodrich were unsuccessful and, before a contract was signed with General Tire of Akron in June 1933, major technological and political changes had transformed the situation.

In November 1931, Du Pont announced that it was launching a new synthetic rubber, a polymer of 2-chlorobutadiene (chloroprene), made from acetylene. It was originally called Duprene, but this was changed to neoprene in 1936 to avoid associating Du Pont with the finished products, which might be of poor quality. The effective marketing strategy of Ernest R Bridgewater, the sales manager of the Rubber Chemicals Division, was the crucial factor in neoprene's commercial success. Faced with low natural rubber prices, he was convinced that neoprene had to be presented as a wholly new material with special properties, not just a new synthetic rubber. To persuade rubber manufacturers to use idle equipment to process neoprene, it had to be made comparable to natural rubber in processing characteristics. Du Pont kept in close contact with the rubber industry and publicised its new product widely, sending salesmen to large potential consumers and publishing material for smaller ones. As John K. Smith has remarked, Du Pont indirectly boosted demand for neoprene "with this well-orchestrated sales campaign aimed at fabricators, engineers and consumers". This strategy enabled Du Pont to sell neoprene for 65 cents a pound in 1939, compared with a natural rubber price of 18 cents a pound.

The successful introduction of neoprene posed a major threat to the Buna copolymers, and IG Farben quickly appreciated the potential of a relatively cheap synthetic rubber that could command a premium price for its special properties, particularly its resistance to oil, solvents and aerial oxidation. That Buna S survived to become the most important synthetic rubber, at least in terms of volume, was the result of an equally dramatic upheaval in German and international politics.

BUNA S IN THE THIRD REICH

Hitler came to power at the end of January 1933, determined to increase Germany's economic self-sufficiency and military preparedness. Synthetic rubber was one of the few fields in which autarky (self-sufficiency) was technologically feasible and would be vital in a motorised *blitzkrieg* if Britain imposed a naval blockade. It also complemented Hitler's interest in motor transport: his people's car (Volkswagen) would glide along the new autobahns on all-German tyres. Complex and energy-intensive, yet compact and clean, with an aura of high-technology, the desired Buna works were the ideal showpiece factories.

By contrast, IG Farben was sceptical about German self-sufficiency in rubber. In June 1933, IG's chief rubber chemist, Erich Konrad, wrote to Fritz ter Meer, the Vorstand member in charge of synthetic rubber, that "synthetic rubber is simply unsuitable for an autarky experiment". IG's negative view of autarky was reinforced by its growing interest in the development of oil-resistant rubbers. Du Pont was demonstrating how it was possible to sell speciality rubbers at a premium, even during an economic depression. When Conti told Konrad in May 1933 that it was testing neoprene, he countered by offering Buna N, the oil-resistance of which had only recently been discovered by IG. IG began small-scale production of Buna N in 1934 and used Buna N, rather than Buna S, in the tests arranged with General Tire. The Akron firm experimented with Buna N in the spring of 1934, but complained that the synthetic rubber was destroying its machinery and was "definitely inferior" to natural rubber. At this stage, IG aimed to satisfy the government's demand for synthetic rubber with Buna N, and sell any excess production as an oil-resistant rubber in competition with neoprene.

After consolidating his power in the summer of 1934, Hitler stressed that he wanted the development of synthetic rubber to be pursued with "an elemental force". When IG Farben failed to match his expectations, he goaded it into action by declaring at the 1935 Nuremburg rally that "the erection of the first [synthetic rubber] factory in Germany ... will be started at once". As he later wrote in a secret memorandum, "from now on there must be no talk of processes not being fully determined and other such excuses".

This pressure put IG in a quandary, as the company had not been able to obtain a licence for neoprene from Du Pont, nor was it ready to scale up the Buna S technology. Buna N had been found in the summer of 1934 to be unsuitable as a tyre rubber, because it could not be blended or repaired with natural rubber. In an attempt to secure a licence for neoprene, Fritz ter Meer travelled to Du Pont in Wilmington, Delaware, in October 1934, but the negotiations ended in deadlock. Tests carried out in November 1935 showed that Buna S was possibly superior to neoprene as a tyre rubber, but the dual benefit of a general-purpose rubber that was also oil-resistant was so irresistible to IG that ter Meer continued to negotiate with Du Pont. The two companies eventually signed a modest agreement in September 1938. Du Pont gave IG a licence to use Du Pont technology to produce butadiene (which was not taken up by IG), but not to make neoprene. The Germans honoured this restriction and, while IG Farben prepared neoprene in experimental quantities during the war, it was never made on an industrial scale.

Under increasing pressure from Hitler, and lacking a neoprene licence from Du Pont, IG Farben was forced at the end of 1935 to choose Buna S as its main synthetic rubber. This haste reduced the quality of Buna S and hindered the development of better synthetic rubbers. The firm attempted to meet the government's demands by building a pilot plant at Schkopau in March 1937 which was scaled upto a small, 24,000 tonnes a year plant in May 1938. The main works at Schkopau were completed in April 1939, and the second factory at Hüls, near Recklinghausen in the Ruhr, started up in August 1940. After a long-running disagreement over the necessity of, and the site for a third factory, IG agreed after the outbreak of war to build it at Rattwitz, near Breslau. The firm unilaterally abandoned the site in July 1940, after the French capitulation, but was forced to make provision for two factories a few months later when Britain failed to surrender. One was erected alongside the Ludwigshafen works, against the advice of the armed forces, on the condition that the fourth plant was located beyond the reach of British bombers. IG selected a site at Auschwitz, about seven kilometres from the concentration camp, but the factory never produced Buna S, despite the extensive use of forced labour. German production of Buna S increased from a mere 2,110 tonnes in 1937 to a wartime peak of 110,569 tonnes in 1943.

To overcome the processing problem, IG developed the thermal degradation (*thermisches Abbau*) process with the co-operation of Conti. This softened Buna S by heating it at a high temperature in air, breaking down the long polymer chains into more easily processed fragments, which also improved some of its physical properties. However it was also a time-consuming and technically tricky process, and the treated rubber had to be used within 24 hours. The introduction of compounds into the polymerisation recipe which "moderate" the growth of the polymer chains and prevent them from becoming too long proved to be a better alternative. A range of moderators or "modifiers" had been developed for sodium polymerisation by the Ludwigshafen laboratory in 1929. The first suitable modifier for Buna S, linoleic acid, was patented during the scaling-up research at Leverkusen in June 1936. Better modifiers, which contained sulphur, were discovered at Leverkusen by Kurt Meisenburg's group (diproxid or "dixie") and Wilhelm Becker's group (long-chain mercaptans, later used in the USA) during 1937. IG preferred the original modifier until the growing shortage of linoleic acid impelled the introduction of sulphur-based modifiers in 1943 (Buna S3). However even Buna S3 required thermal degradation before processing.

IG Farben's attitude towards the synthetic rubber programme was ambivalent. Although it was not keen on the idea of complete autarky in rubber, it saw the programme as a means of underwriting its plans for the development of acetylene-based organic chemicals. IG's technical director, Fritz ter Meer, later commented that acetylene was "a new chemical feedstock, useful in many types of chemical synthesis, and in a number of our laboratories we specially directed research in the acetylene field". Between 1928 and 1944, Walter Reppe of Ludwigshafen discovered entirely new acetylene reactions, which promised to make it the starting-point for numerous new chemicals. If the synthetic rubber programme assisted the development of acetylene chemistry by funding research and the construction of acetylene works, it also threatened to overwhelm it. As long as the relatively unprofitable production of Buna S continued, it drained IG's reserves of capital and scientific personnel, and crowded the other acetylene-based processes out of the new factories.

TRANSFER OF BUNA S TECHNOLOGY TO THE USA

The American rubber industry shared the German industry's concern about British attempts to push up the price of natural rubber and being cut off from the Far East in wartime. However the development of synthetic rubber in the USA followed a quite different path. Before the British naval blockade during the First World War forced the USA to manufacture dyes and organic chemicals hitherto imported from Germany, the Americans lacked a dyestuffs industry of any size. In terms of size, international influence and technological innovation, the petroleum corporations were the American equivalents of the German dye companies. Indeed IG Farben was loosely modelled on the Standard Oil Trust, and Standard Oil of New Jersey was one of the few firms openly admired by the leaders of the German combine. Furthermore the American rubber companies were much larger and more innovative than their German counterparts. Continental Gummiwerke was even partly owned by Goodrich between 1920 and 1929. Du Pont was the only chemical corporation on an equal footing with the petroleum and rubber companies before Pearl Harbor.

The importance of the American petroleum and rubber companies was a result of the early and rapid development of the American motor-car industry to a size unmatched anywhere else. In 1937, the USA had a population of 129 million, about twice the German population of 68 million, yet it manufactured nearly four million motor-cars, compared with only 269,000 in Germany; the USA had 23 private motor-cars registered for every one in Germany. Not surprisingly, the USA produced 60 million tonnes of motor spirit in 1937 and consumed 552,300 tonnes of natural rubber, exactly half of total world consumption. By contrast, Germany only produced 1.3 million tonnes of motor spirit (including some synthetic) and consumed about 80,000 tonnes of natural rubber. Even Du Pont partly owed its rise to pre-eminence to its close, and profitable, relationship with General Motors. The big four rubber companies (Goodyear, Goodrich, Firestone and US Rubber) dominated the development of synthetic rubber in America, with the marginal exception of Du Pont's neoprene. Naturally this ensured that far greater attention would be taken of the needs of the rubber processors and the final consumers, but, perhaps surprisingly, the Americans were quickly able to match IG in terms of chemical expertise. How did this come about and how was the necessary "know-how" transferred from Germany to the USA?

The giant oil company Standard Oil of New Jersey was an important go-between in the transatlantic transfer of synthetic rubber technology. The early 1920s saw an upsurge of research on petroleum refining, especially the cracking of heavier oils into the gasoline used in motor-cars, based on the earlier research of William M. Burton at Standard Oil of Indiana, a smaller fragment of the former oil trust. Standard Oil of New Jersey was not at the cutting edge of this technology and was also concerned about declining reserves in the traditional American oilfields. IG Farben was developing the Bergius process for converting coal into gasoline, which could be used for heavy oils or shale oil, but its development was expensive and IG was running out of funds. With its long experience of petroleum refining, Jersey Standard could assist with the development of the Bergius process and also help to distribute the coal-based gasoline in Germany.

The first contact was made by BASF in 1925, just before the formation of IG, and the two sides reached a limited agreement in August 1927. In return for the rights to the oil-from-coal process in the USA and half of the royalties from third parties, Standard agreed to fund research into the process at its new Baton Rouge laboratories. Encouraged by its early results with the conversion of heavy oils, Jersey Standard then agreed, in November 1929, to form the Standard-IG Company (four-fifths of which was owned by the American firm) to hold all the rights to IG's oil-from-coal patents outside Germany. The German combine was compensated with two per cent of Standard's common stock. Standard also agreed in a further covenant to refrain from entering the chemical field, while IG undertook to keep out of petroleum refining. Any process patented by IG in the petroleum field would therefore be offered to Jersey Standard on reasonable terms, and vice-versa.

Goodrich and Goodyear, the two companies most interested in synthetic rubber, were willing to adopt IG's technology, given reasonable terms. Jersey Standard had approached Goodrich, on IG's behalf, in 1932 to seek help with the processing of the copolymers. The two companies were unable to reach agreement, and the contract for the testing was given to the smaller General Tire. Goodrich were still interested in the Buna rubbers and Waldo Semon, the leader of its synthetic rubber team, visited Germany in the summer of 1937, in a vain attempt to exchange Goodrich's superior plasticised PVC technology for IG's Buna "know-how". Goodyear was also contacted by Standard in 1933, and expressed an interest in developing the Buna copolymers. On the basis of the scanty information they obtained, the Goodyear chemists made significant advances with the polymerisation process. Lorin Sebrell of Goodyear also made a pilgrimage to IG's Frankfurt headquarters in an attempt to reach an agreement with the German firm, but to no avail. IG Farben's experts could not accept that the American rubber companies were capable of making their own synthetic rubber, and even accused Goodyear of making its samples (including a complete tyre) from pirated German Buna S. Even Jersey Standard, IG's legal partner in the world-wide development of synthetic rubber, was able to learn little about the large-scale manufacture of Buna S and Buna N, in contrast to the free exchange of information in the oil-from-coal field. Irked by IG Farben's disdain, Goodrich and Goodyear independently pressed ahead with the development of their own copolymer rubbers. To circumvent IG's patents, they made a large number of experimental copolymers that did not use styrene or acrylonitrile; Goodrich called them non-infringing rubbers or "nirubs".

Why did IG fail to respond favourably to Goodrich's and Goodyear's overtures? Most commentators have assumed that the German government (and IG) aimed to prevent the USA from developing Buna S. Not surprisingly, there were German government restrictions on the dissemination of military-related technology. On the question of transfer of Buna technology to the USA, the Reich Ministry of Economics took a straightforward position, at least up to early 1939. The ministry realised that IG Farben was likely to face commercial and technological competition from American companies if it did not license the Buna technology, and hoped that this transfer might reduce American hostility towards Nazi Germany. The ministry officials and the army high command (OKW) were therefore willing for IG to reach an agreement with the American firms. However there were two conditions: the final agreement would need the approval of the Reich Ministry of Economics (and indirectly OKW) and the work on the German Buna factories had to come first. It is also probable that the very latest technology would have been withheld, though a number of hitherto secret patent applications were released with the permission of OKW in October 1938. Of course there was no guarantee that the ministry would have granted permission, but it was willing to accept that Du Pont's neoprene and the two-step butadiene process based on Du Pont's technology could be superior to IG's technology, and would have been unlikely to object if IG had exchanged its processes for American "know-how".

With its usual arrogance, however, the German firm believed that no other firm could develop a good copolymer without its cooperation. At the same time, IG feared the entry of other firms into the synthetic rubber field. It was thus reluctant to divulge its precious "know-how" without control over its use and substantial compensation. IG refused to give information about Buna to the French Vichy government, and resisted attempts by the Italian synthetic rubber company to obtain the formula for Buna S3, despite the agreement of the German government to the transfer. Du Pont, with its neoprene patents, was the one American firm that could offer IG technology of equal value and IG attempted to use its Buna patents to reach a wide-ranging agreement with Du Pont. The impetuous development of Buna copolymers, particularly the oil-resistant Buna N, by Jersey Standard and/or the rubber companies might have perturbed Du Pont and made an agreement less likely.

In any event, the outbreak of war in Europe in September 1939 made it difficult to maintain the agreements between Jersey Standard and IG Farben. While the two companies did not expect war to break out between Germany and the USA, their cooperation posed problems for Standard's British and French associates; it was also becoming a political embarrassment for both sides. IG Farben wanted to be free to sell the Buna technology to the Italians and Russians, countries within Jersey Standard's sphere under the original agreements, while Standard was eager to pursue the development of synthetic rubber in America even without IG's assistance. After three days of complex negotiations in The Hague, Jersey Standard's Frank Howard and IG's Fritz Ringer signed an agreement, on 25 September 1939, which dissolved the original arrangement and gave Jersey Standard the right to develop Buna S in the USA, the British Empire and the French Empire in return for giving up their joint rights elsewhere. IG did not lose much by this agreement. Jersey Standard already controlled the US Buna patents and IG's ownership of the British and French rights was compromised by the outbreak of war. Furthermore IG Farben later declined to hand over the technical "know-how" on Buna S manufacture - in particular information about IG's full-scale production - to Jersey Standard to avoid a potential collision with the German government.

Nevertheless, armed with the full patent rights and the results of its own research, Jersey Standard suggested to the major rubber companies that they all cooperate to develop Buna S by forming a joint company; a proposal not dissimilar to ter Meer's earlier plan. This was acceptable to Firestone and US Rubber, who had not carried out much synthetic rubber research, but Goodrich and Goodyear declined the offer. In June 1940, Goodrich defied Standard's attempts to coordinate the development of synthetic rubber by launching Ameripol (American polymer), a copolymer of butadiene and methyl methacrylate. For a thirty per cent premium over the normal price, patriotic citizens and companies could buy Liberty tyres, made from a blend of natural rubber and Ameripol.

At this stage, however, the rubber companies and Jersey Standard were more interested in the highly profitable speciality rubbers, such as Buna N, rather than tyre rubbers. German Buna N was first exported to the USA in the third quarter of 1937. It was able to gain a toehold in the American market after a major explosion in Du Pont's Deepwater, New Jersey, plant in January 1938 disrupted neoprene production for several months. By June, IG's agent had gained orders for 20 tonnes of Buna N, which had been renamed Perbunan, apparently to circumvent a government export ban on Buna N. American stocks of Buna N soon ran out in the autumn of 1939, and the rubber companies were keen to make it themselves. By the end of 1940, Goodrich and Goodyear had independently erected small Buna N plants and sold it without a licence from Standard.

After waiting almost a year, Standard took the two rubber companies to court in October 1941 for alleged infringement of its patent rights. To protect itself, Goodrich had already carried out experiments to show that the Buna copolymers could not be prepared from the patents, as required by law. The legal problems and the divide between the three rival groups (Standard-Firestone-US Rubber; Goodrich-Phillips; Goodyear-Shell-Dow) were overcome by the Japanese attack on Pearl Harbour in December 1941. Within a week of that event, Jersey Standard and the big four rubber companies had signed a patent- and information-sharing agreement under the auspices of the government-owned Rubber Reserve Company.

THE AMERICAN SYNTHETIC RUBBER PROGRAMME

After several changes of policy, and the appointment of a high-powered commission of inquiry (the Baruch Committee) by President Roosevelt, a joint production and research programme, supervised by a Rubber Director, was inaugurated in October 1942. The companies in the programme agreed upon a "mutual" recipe for GR-S (government rubber-styrene), which was similar to the German recipe for Buna S. The crucial difference was the replacement of linoleic acid by one of the sulphur compounds patented by IG in 1937. This was usually called "OEI": one essential ingredient. The Americans were enthusiastic about modifiers, because they produced an easily processible rubber which could be worked on existing machinery.

The expansion from an American synthetic rubber industry with an output of 231 tonnes of Buna S in 1941 to one that was producing 70,000 tonnes a month of GR-S in the spring of 1945 was a remarkable achievement. As with other American wartime industries, the reserve capacity of the US economy and the sheer scale of production were astounding. There were four striking differences from IG's operations: the polymerisation plants were physically separate from the monomer plants; they were operated by the rubber industry; there was a large number of plants scattered across the country; and the major monomer, butadiene, was made from grain alcohol or petroleum, not coal-based acetylene.

A total of 15 polymer plants were planned; all were operating by December 1943. The butadiene-from-alcohol plants started up in the first half of 1943, and were soon producing at up to twice their rated capacity. The butadiene-from-petroleum plants were slow to come on stream, for a number of reasons, and the alcohol-based plants carried the burden for most of 1944. By 1945, however, as the petroleum-based plants came into full operation, GR-S production surged ahead of demand. Total GR-S production in 1945 was 730,000 tonnes, six and a half times German Buna S output at its peak. Smaller amounts of speciality rubbers - Buna N (GR-A), neoprene (GR-C), butyl rubber (GR-I) and Thiokol (GR-P) - were also produced during the war. The new Du Pont neoprene factory at Louisville, Kentucky expanded production from 10,000 to 60,000 tonnes a year with government help, and was bought back by Du Pont in 1949.

From the outset the programme's leaders realised that a research and development programme would be necessary to solve the existing and potential problems surrounding GR-S manufacture. For instance, OEI rapidly disappeared during the polymerisation process, thereby producing an unsatisfactory rubber. R R Williams of Bell Telephone Laboratories (BTL) was commissioned to set up and supervise this programme, which was centred on the rubber industry, the National Bureau of Standards, BTL and the major universities. Williams drew on his experience with rubber and other polymers at BTL when he laid down the guidelines for the research programme at the end of 1942. The chemists at BTL believed that the quality of a polymer was crucial to its performance, and the study of a polymer's structure was the key to quality improvement. The problem of supervising numerous research groups, whose work often overlapped, was solved by adopting a "hands-off" approach which allowed the groups largely to direct their own research. This liberal philosophy was shared by IG Farben, which had faced similar research management problems. It reduced conflict between the central administration and the groups, but the trade-off was increased duplication between different research laboratories and the loss of a strong sense of direction.

The first year of the research and development programme was naturally dominated by the need to solve existing problems. The difficulty with the disappearing modifier was overcome by adding the modifier to the polymerisation cycle gradually, and replacing OEI with another group of sulphur compounds called tertiary mercaptans. A puzzling and unpredictable pause at the beginning of polymerisation was traced to polyunsaturated acids in the Ivory soap used as the emulsifier. It was important to keep the butadiene-styrene ratio of GR-S constant, and William O Baker of BTL developed a new method of measuring the styrene content of GR-S.

Once the teething problems were overcome, the researchers were able to concentrate on the chemistry of the polymerisation and the structure of GR-S. A central role was played by chemical analysis, which was essential to quality control and kinetic studies. Through his analytical research, Piet Kolthoff at the University of Minnesota was drawn into the study of the various factors which affected the rate of polymerisation. Frank Mayo's basic research group at US Rubber converted an old silk mill in Passaic, New Jersey into a major centre for the study of copolymerization. W D Harkins, a retired professor at the University of Chicago, developed a detailed model for emulsion polymerisation which dominated the field for many years. At BTL, Baker revealed that harmful gel particles were formed during the polymerisation. The Dutch chemical physicist Peter Debye, who fled Germany in 1940 and joined Cornell University, demonstrated how the visible light scattered by polymers in solution could be used to measure their molecular weight. The novel techniques of infra-red and ultra-violet spectrophotometry were used to study the structure of GR-S.

The American research programme did not make any outstanding technological breakthroughs before the war's end but, combined with stringent quality control, it created a GR-S that was superior to IG's vaunted Buna S. The strength of American chemical research was soon confirmed by the rapid American exploitation of IG's breakthrough in redox polymerisation, based on information brought back by American technical intelligence units after the German defeat in 1945.

THE POSTWAR PERIOD: 1945-1953

Because of the threat posed by the Soviet Union, the Americans felt it was necessary on the grounds of national security to maintain a significant synthetic rubber industry after the war ended. Synthetic rubber production was naturally cut back, but it was nevertheless maintained at 300,000 to 400,000 long tonnes of GR-S a year. For the same reason, the plants closed down were mothballed rather than scrapped, or sold on the condition that they could be reactivated for synthetic rubber production at short notice. Government control of the plants and rubber inventories was retained, but the Rubber Act of 1948 called for the disposal of the plants by 1950.

The main aim of the immediate postwar research was the development of a rubber that could compete with natural rubber on both quality and price. In October 1948, the Rubber Reserve Company announced the introduction of "cold rubber", a superior form of GR-S, which was an excellent rubber for automobile tyre treads. This new rubber was made possible by a new catalyst system which permitted the polymerisation of butadiene and styrene at $5\,^{\circ}C$ ($41\,^{\circ}F$) instead of the standard $50\,^{\circ}C$ ($122\,^{\circ}F$), hence the name "cold" rubber. The drop in polymerisation temperature produced a rubber that was strong but without the harmful gel found in "hot" GR-S. However, cold rubber was not simply the old GR-S with a lower polymerisation temperature. Its manufacture incorporated other advances, such as tertiary mercaptans, a new kind of soap for the emulsion, and greatly improved carbon blacks (for tyre reinforcement) developed by Phillips Petroleum.

Cold rubber was the result of a decade's research by the four large rubber companies and Phillips Petroleum. It was originally developed by IG Farben during the Second World War, but never taken to the industrial scale. The "redox" recipe patented by Heino Logemann of Leverkusen in 1943 was improved by Piet Kolthoff in 1946. The "Custom" recipe adopted by Rubber Reserve in 1948 for the standard production of cold rubber was based largely on research by William Reynolds and Charles Fryling at Phillips Petroleum. In 1950, cold rubber already accounted for thirty-eight per cent of all GR-S production. It was the equal of natural rubber for automobile tyres, but was also difficult to process and no cheaper than the old GR-S.

After the Communist invasion of South Korea in June 1950, the focus of the programme switched from quality to quantity. Even before the war began, the price of natural rubber had been spiralling upward, partly because of heavy buying by the Soviet Union. It reached a peak of 73 cents/lb compared with the fixed price of 18.5 cents/lb for GR-S. By September, the synthetic rubber industry had been completely reactivated. GR-S production in 1951 was 697,000 long tonnes, almost double the previous year's total. To cover the additional cost of the alcohol-based butadiene from the standby plants, the government raised the price of GR-S to 24.5 cents/lb.

Oil-extended rubber (a 4:1 mixture of cold rubber and selected mineral oils) was introduced independently in 1951 by General Tire and Goodyear as a way of increasing the volume of rubber production. The addition of mineral oil also made cold rubber easier to process and cheaper. The output of GR-S could be increased in other ways. The Government Laboratories in Akron (founded in 1944) and Goodrich collaborated on the development of a "rapid" GR-S which used a catalyst system created for cold rubber to reduce the polymerisation period at 50°C. This doubled the rate of production and hence greatly increased plant capacity.

While cold rubber was adequate for automobile tyres, it was not suitable for an all-synthetic large truck tyre (or the similar aircraft tyres). Truck tyres usually operate at high temperatures under heavy loads for long periods. GR-S tyres suffered excessive heat build-up while running, and this problem was accentuated by the synthetic rubber's poor strength at high temperatures. Goodrich had hoped to solve this problem during 1942, but it remained obdurate.

In March 1950, Goodrich converted part of its pioneering research section (unconnected with the government programme) into the "American Rubber Team", directed by Waldo Semon, with a mission to develop a synthetic rubber that could be used to make heavy-duty tyres. The group, fifty strong at its peak, failed in its main objective, but several advances in rubber technology were made. Carl Marvel's group at the University of Illinois attempted to solve this problem by replacing styrene with compounds related to benzalacetophenone, but no significant improvement in performance was obtained, given the higher cost of the new polymers. This project initiated Marvel's research on heat-resistant fibres in the 1960s and 1970s, which resulted in the commercialisation of polybenzimidazole (PBI) fibre in 1983.

THE END OF THE AMERICAN RUBBER PROGRAMME: 1953-1955

Even when the Korean War faded into the background in 1952 and 1953, GR-S production remained high at an annual level of 600,000 long tonnes. The superior properties of cold rubber and the cheapness of oil-extended rubber (thirty-one per cent of cold rubber and fifteen per cent of all GR-S in 1953) made GR-S more competitive with natural rubber. This increased the pressure to put the synthetic rubber industry into the private sector. The 1950 deadline in the 1948 Rubber Act had been extended to 1952, and then to 1954, because of the war.

The Reconstruction Finance Corporation (RFC) presented its disposal plan to Congress in 1953. The RFC and Congress were anxious to prevent the big four rubber companies from forming an oligarchy in synthetic rubber production, and to ensure that the plants would be available in a national emergency for GR-S production. Congress rejected the RFC's plea to be put in charge of the disposal operation, and the 1953 Rubber Producing Facilities Disposal Act created an independent commission for this purpose. The Attorney General was instructed to monitor the disposal process to prevent violations of the anti-trust laws. The twenty-nine plants were to be widely advertised for sale, and the sale contracts would contain a clause that the plants would be available for emergency use for ten years after the sale.

The disposal commission, headed by Leslie Rounds, a former vice-president of the Federal Reserve Bank, met for the first time in November 1953, and the advertisements appeared eight days later. The deadline for bids was 27 May 1954, and by that date seventy-five proposals had been received from thirty-five bidders. The number of bids per plant ranged from eleven for the Los Angeles styrene plant to none at all for the Institute, West Virginia, GR-S plant and the two butadiene-from-alcohol plants. The commission then negotiated with the bidders (the details of the bids were kept secret) to raise the prices bid and to make them more even across the board. It was determined not to sell a plant unless a fair price could be obtained. The first contract was signed on 16 December and the final one on 27 December, the last day permitted by the Act. Twenty-four of the twenty-nine plants had been sold; the bid for the Baytown, Texas, GR-S plant was rejected as too low. The sale of the plants, not including inventories, raised just over $260 million. Congress approved the sales in March 1955, and the transfers took place in late April. The Baytown plant was handed over to United Carbon on 15 July 1955, following a second round of bids, and the US government was out of the synthetic rubber business.

This period also witnessed a major turnaround in synthetic rubber research. The problem of the large truck tyre was finally solved by synthesising natural rubber itself. Firestone started a special research effort in 1952, determined to succeed where Goodrich's American Rubber Team had failed. Its research director, Frederick Stavely, decided to concentrate on the polymerisation of isoprene (the monomer of natural rubber) with the active metal, lithium. This line of research succeeded in making a polyisoprene rubber in early 1953, and by September a pilot plant was in operation. Polyisoprene can have three arrangements in three-dimensional space: all-*trans*, all-*cis*, or a mixture of the two. Natural rubber owes its superior properties to its all-*cis* arrangement. Firestone's original polyisoprene had a relatively low *cis* content, but Stavely's team was soon able to produce "coral rubber" with a higher *cis* content.

Unaware of Firestone's success, Goodrich heard of a new catalyst system, which had been developed in West Germany by Karl Ziegler to polymerise ethylene. Goodrich-Gulf (a joint subsidiary of Goodrich and Gulf Petroleum) obtained a licence to use this catalyst from Ziegler in the summer of 1954. A young chemist, Samuel Horne, was detailed to make a copolymer of ethylene and isoprene, and unwittingly prepared an analogue of natural rubber. A public announcement of the new process was made by Goodrich-Gulf on 2 December 1954. Commercial production of Ameripol SN (Synthetic Natural) began in 1958.

Within a year, three other companies had independently developed processes for the manufacture of synthetic "natural rubber" and the similar all-*cis* polybutadiene. At Phillips Petroleum, Robert Zelinski prepared an all-*cis* polyisoprene in December 1954, only to learn of Horne's work. He went on to develop all-*cis* polybutadiene, which was commercialised by Phillips in 1959. The chemists at Goodyear, who did not know the details of the Goodrich process, also succeeded in producing an analogue of natural rubber with a Ziegler catalyst in the summer of 1955. Goodyear started production of Budene (*cis*-polybutadiene) at the end of 1961 and Natsyn synthetic "natural rubber" followed four months later. Finally, Lee Porter of Shell Chemicals used a lithium-cobalt catalyst to produce a polymer like Firestone's coral rubber. "Shell Isoprene Rubber" also went into industrial production in 1959.

SYNTHETIC RUBBER SINCE 1955

Total SBR (styrene-butadiene rubber) capacity in the United States expanded from 890,000 long tonnes in 1955 to 1.4 million long tonnes in 1957. The Goodyear plant in Houston, Texas, alone had a capacity of 200,000 long tonnes in 1957. American SBR production (excluding latex) topped 1.5 million long tonnes in 1975, but the industry then went into a decline because the oil crisis made synthetic rubber more expensive relative to natural rubber. Except for a brief rally in 1977-1979, when production almost touched 1.4 million long tonnes, American SBR production has consistently hovered around 900,000 long tonnes.

Since the government plants were sold in 1955, there has been only one major technological innovation in the synthetic rubber field; the ethylene-propylene copolymers and terpolymers. This completely new class of synthetic rubber was made possible by Ziegler's new catalyst system, which was improved in the mid-1950s by Giulio Natta of Milan Polytechnic. Whereas butadiene can be polymerised to a rubbery material on its own, neither ethylene nor propylene can be made into rubbers by themselves. The first ethylene-propylene rubber was commercialised by Jersey Standard in 1960, and independently by the Italian chemical company Montecatini in 1961. The original copolymer could not, however, be easily vulcanised. The superior terpolymer, with a small amount of a third monomer that permits cross-linking, was announced by Du Pont in 1961 and a semi-industrial plant started up in July of that year. Dunlop Canada introduced their own terpolymer, with a different termonomer, in 1962. Du Pont opened its full-scale plant in 1963, but Dunlop Canada and its partner, Hercules, never got beyond the pilot stage. By 1967, Uniroyal, Copolymer Corporation, and Jersey Standard were producing ethylene-propylene rubber using a termonomer patented by Du Pont.

The hallmark of the post-1955 industry has been technology transfer rather than innovation. The major West German synthetic rubber producer Chemische Werke Hüls (a former joint subsidiary of IG Farben and a coal firm), decided to abandon its homegrown but obsolescent technology and purchase a GR-S plant from Firestone in 1954. Two years later it acquired the technology from the Houdry Process Corporation to make butadiene from petroleum. Synthetic rubber production began in Britain and Italy in 1957, in France and Japan two years later, and in Brazil (the original home of natural rubber) in 1962. By the early 1970s, Japan had established herself as a major producer of not only SBR but also all-*cis* polybutadiene and polyisoprene. Japanese production of SBR is still only about two-thirds of the level of American production, but the gap is rapidly narrowing.

NOTE

This paper is an abridged version of Peter J. T. Morris, 'Transatlantic Transfer of Buna S Synthetic Rubber Technology, 1932-45' in David J. Jeremy (ed.), 'The Transfer of International Technology: Europe, Japan and the USA in the Twentieth Century' (Aldershot, 1992), augmented by material from Peter J. T. Morris, 'The American Synthetic Rubber Research Program' (Philadelphia, 1989). Reproduced by kind permission of Edward Elgar Publishing Limited and the University of Pennsylvania Press.

SELECT BIBLIOGRAPHY

Austin Coates, 'The Commerce in Rubber: The First 250 Years', Oxford University Press, Singapore, 1987.

James D. D'Ianni, 'Fun and Frustrations with Synthetic Rubber', Rubber Chemistry and Technology, 1977, 50, G67-G77.

Peter Hayes, 'Industry and Ideology: IG Farben in the Nazi Era', Cambridge University Press, Cambridge, 1987.

Vernon Herbert and Attilio Bisio, 'Synthetic Rubber: A Project That Had to Succeed', Greenwood Press, Westport, Connecticut, 1985.

David A. Hounshell and John K Smith, 'Science and Corporate Strategy: Du Pont R & D, 1902-1980', Cambridge University Press, Cambridge, 1988.

Frank A. Howard, 'Buna Rubber: The Birth of an Industry', Van Nostrand, New York, 1947.

Paul Kränzlein, 'Chemie im Revier: Hüls', Econ Verlag, Düsseldorf and Vienna, 1980.

Bettina Löser, 'Der Einfluss der Arbeiten zur Strukturaufklärung und Synthese der Kautschuks auf die Herausbildung der makromolekularen Chemie', Karl-Marx-Universität, Leipzig, PhD thesis, 1983.

Frank M. McMillan, 'The Chain Straighteners. Fruitful Innovation: The Discovery of Linear and Stereoregular Synthetic Polymers', Macmillan, London, 1979.

Herbert Morawetz, 'Polymers: The Origins and Growth of a Science', Wiley-Interscience, New York, 1985.

Peter J.T. Morris, 'The Development of Acetylene Chemistry and Synthetic Rubber by IG Farbenindustrie AG, 1926-1945', Oxford University DPhil thesis, 1982.

Peter J.T. Morris, 'Polymer Pioneers', Center for the History of Chemistry Publication No 5, Philadelphia, 1986.

Maurice Morton, 'History of Synthetic Rubber' in 'History of Polymer Science and Technology', ed. Raymond B Seymour, Marcel Dekker, New York and Basel, 1982, pp 225-239.

Maurice Morton, 'Rubber Enters the Polymer Age', Rubber Chemistry and Technology, 1985, 58, G75-G90.

Gottfried Plumpe, 'Die I.G. Farbenindustrie AG: Wirtschaft, Technik, Politik, 1904-1945', Duncker & Humblot, Berlin, 1990.

Davis R.B. Ross, 'Patents and Bureaucrats: US Synthetic Rubber Developments before Pearl Harbour', in 'Business and Government', eds. Joseph R Frese, S J, and Jacob Judd, Sleepy Hollow Press, Tarrytown, New York, 1985, pp 119-55.

Robert Solo, 'Synthetic Rubber: A Case Study in Technological Development under Government Direction', Study No 18 for the Sub-Committee on Patents, Trademarks and Copyright, Committee on the Judiciary, US Senate, 85th Congress, 2d session, 1959, Committee Print, 93, reprinted as 'Across the High Technology Threshold: The Case of Synthetic Rubber', Norwood Editions, Norwood, Pénnsylvania, 1980.

Peter H. Spitz, 'Petrochemicals: The Rise of an Industry', Wiley-Interscience, New York, 1988.

William M. Tuttle, Jr., 'The Birth of an Industry: The Synthetic Rubber "Mess" in World War II', Technology and Culture, 1981, 22, pp 35-67.

G.S. Whitby, C.C. Davis and R.F.Dunbrook (eds.), 'Synthetic Rubber', Wiley, New York and London, 1954.

Franz I. Wünsch, 'Das Werk Hüls: Geschichte der Chemische Werke Hüls AG in Marl, 1939-1949', Tradition, 1964, 9, pp 70-9.

Polythene: The Early Years

G. D. Wilson

FORMERLY ICI PETROCHEMICALS AND PLASTICS DIVISION, WELWYN
GARDEN CITY, HERTFORDSHIRE, UK

Introduction

As we look back from the position today, where polyethylene production is a
worldwide industry with a total manufacturing capability in excess of 30 million
tons, to that day in March 1933, when a minute amount of a white waxy solid was
found on the walls of an experimental high pressure bomb, it is difficult to
appreciate what the world must have been like without the benefits of this versatile
material. No plastic bags, no plastic greenhouses or growing tunnels, no fertiliser
sacks, no plastic washing up bowls - the list is endless. Would today's position
have been achieved without the demand generated for products for high frequency
cable applications, and especially for use in RADAR equipment, following the
outbreak of the Second World War? We will never know the answer to that
question, but it is certainly true that the far-sighted views of a group of ICI
scientists working in the Alkali Division of that company in the early and middle
1930s led the way to the discovery of an entirely new range of products and to a
new way of life for humanity.

The Pre War Years

Much has already been written about the initial discovery, but it is not always
appreciated that the 1933 work was not able to be repeated until December 1935,
when 8 grams of the material were produced in a small scale laboratory reactor.
In the intervening period, the first attempt at a repeat of the original work resulted
in an explosive decomposition of the contained ethylene, leading to the production
of a mixture of hydrogen and methane and the deposition of amorphous carbon
within the vessel. The decomposition resulted in the bursting of joints, tubes and
gauges of the apparatus, and it was decided to discontinue the experiments until
more suitable apparatus had been designed.

It may be asked why ICI was so single minded in pursuing this reaction.
However, prior to the attempt to react ethylene and benzaldehyde at a pressure of
1900 atmospheres and a temperature of 170°C in March 1933, some 50
experimental high pressure reactions involving different gaseous mixtures had been
attempted over the previous 15 months, with disappointing results. The production
of that minute amount of waxy solid in the 1933 work was the catalyst for further

work on this particular reaction.

During the period between the end of December 1935 and end May 1936, sufficient work had been carried out on the polymerisation of ethylene in small batch bombs to enable the effect on the reaction of pressure, temperature, gas purity, use of different catalyst systems and of different surfaces, to be more clearly understood. In the course of this work it was found possible to achieve reaction under a number of different conditions, and in particular that the higher the pressure to which the reactor was filled, the lower was the temperature required for reaction to take place. Actual pressures ranged between 3,000 and 500 atmospheres, and temperatures between 150 and 300°C. It was also reported that two major difficulties had been experienced with the reaction, neither of which had yet been overcome. These were:

    1)   · That the yields obtained were very variable and

    2)   That an explosive decomposition to carbon and hydrogen accompanied by a large rise in pressure and temperature is liable to occur - especially at higher reaction pressures.

Evaluation of the products made during this initial series of experiments showed that molecular weight varied with the pressure under which the polymer was formed, ranging from around 2,000 when reaction was carried out at 500 atmospheres to 12,000 - 20,000 at 3,000 atmospheres. At that stage it did not appear to be temperature dependent. In addition to determining molecular weights, a great deal of progress was made in the study of the crystal structure of the polymer, and in measuring its electrical properties, softening point and chemical inertness. Tests were also carried out to determine its ability to produce threads and filaments, to fill simple moulds and to make thin films.

Based upon the results of these tests the following suggestions were made for possible applications:

1)   Waterproof and Protective Coatings

    a)   The impregnation of textiles, paper, leather etc.

    b)   The use of films for wrapping or lining packages.

2)   Adhesive

    a)   Safety Glass.

    b)   Sealing medium for carboys and acid bottles.

    c)   High vacuum sealing medium.

3)   Electrical uses, in particular for various types of insulation.

4)   Textiles, in particular for the spinning and weaving of threads.

5)   Polishes.

6)   Lubricating oils, involving the addition of the polymer to make **heavy** duty lubricants.

7)   Starting material for synthesis of other products.

At that stage in 1936 there was no clear understanding about the mechanism of the reaction, nor of the causes of the relatively frequent explosive decompositions of the reactor contents to carbon and hydrogen, nor of the role played by trace amounts of oxygen, which was present as a contaminant in the ethylene, in initiating the polymerisation reaction. Indeed, at that stage it was considered that "the polymerisation process was a chain reaction starting with some specially activated ethylene molecules which probably needed a rather large energy of activation. These grew as a result of collisions with other ethylene molecules in the gas phase, the activation energy for this chain mechanism being very much smaller than for the formation of starting molecules. Some stopping mechanism must also be postulated and may possibly involve specially oriented collisions, or collisions with particular molecules, or the migration of a hydrogen atom along the polymer chain to form a normal double bond and so cause deactivation." However, in spite of all these gaps in the knowledge about the reaction mechanism, a decision was taken to proceed with the scaling up of the process.

All work up to June 1936 had been carried out in batch bombs with a capacity of 75 cc heated in a furnace, or in a 150 cc bomb fitted with an internal heating element. Pressure was raised in the bomb by a two stage compression system involving a first stage mercury compressor developed by Professor Michels, of Amsterdam University, which delivered ethylene at a pressure of around 800 atmospheres to the reactor. This gas was then further compressed to a pressure of up to 3,000 atmospheres by a hand operated hydraulic oil pump feeding compressed oil to a mercury piston seal within the reactor space. Ethylene, made by the dehydrogenation of ethanol by phosphoric acid, was obtained in cylinders from the British Oxygen Company.

It was now proposed that ethylene should be supplied from a small plant decomposing ethanol catalytically over alumina, and compressed in a two stage process. The first stage would involve a primary compression system, of a design already in existence, for delivery up to about 200 atmospheres into ethylene cylinders, whilst the second would use an entirely novel design of continuous mercury compressor for delivery at pressures up to 3,000 atmospheres. Such a machine, with a maximum capacity of 5 m^3/hr, was ordered from Dikkers and Company in Holland and manufactured under the direction of Michels, who had already developed a suitable design.

For the reaction vessel, an 8 litre vessel with internal heaters had already been designed by W R D Manning, a senior ICI design engineer, and this design formed the basis of the 9 litre vessel, fitted with an internal stirrer, which was eventually installed in the plant.

A preliminary estimate of the costs of production indicated that at the 50 tonnes per year scale, a total cost of around two shillings per pound (10p in today's currency) could be anticipated.

By mid 1936 a trade name, ALKETH, had been chosen for the new product and provisional patent specifications had been lodged covering the manufacturing process, a mechanism for the control of molecular weight and for a number of product applications.

By December 1936 1.13 kg of the product had been made on a newly designed 750 cc batch reactor, and at the first meeting of the "Alketh" steering committee it was agreed that samples from the first batch of product should be sent to ICI Nobel Division for moulding, rolling and cold extrusion evaluation, and to Metropolitan Vickers for electrical assessment. As work progressed on this vessel, it was also agreed that a small sample should be sent to Du Pont in America, with whom ICI had a technical collaboration arrangement, and to the Shirley Institute in the UK for thread testing.

In March 1937 the 750 cc batch vessel was uprated to 940 cc, and a refinement of the cost of production calculation, based on work with this vessel, suggested a figure of two shillings and six pence per pound (12.5p). By May 1937 the stock of Alketh stood at 780 grams of product with a molecular weight (MW) of 8,000, 690 grams with an MW of 3,600 and 210 grams with an MW of 6,000.

The first product test results obtained by Metropolitan Vickers indicated a power factor of 0.0016 at 50 cycles, and 0.0015 at 1 million cycles. The breakdown voltage tests gave 1,000 to 1,500 volts per mm. At the same time, work within ICI had demonstrated that transparent films could be made with thicknesses up to 0.005 ins which, it was suggested, could possibly be of use in covering aircraft wings because of the product's "high water resistance and good low temperature properties".

The 9 Litre Reaction Vessel Pilot Plant System

In November 1937, just over 1½ years following the decision to scale up the work on the development of the process, the 9 litre reaction vessel was started up on the Wallerscote site.

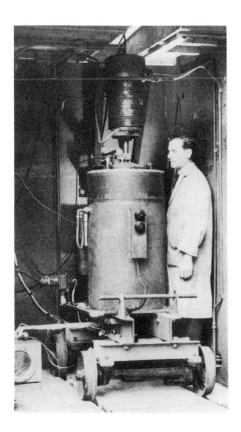

Fig.9. 9 litre reaction vessel. Courtesy of ICI.

The first polymer was manufactured at a pressure approaching 900 atmospheres and a temperature of about 200°C, with a molecular weight of about 10,000. During the succeeding 3 months a range of products were made, with molecular weights ranging between 11,000 and 27,000 at rates up to 1.6 kg/hr.

A significant difference between the 9 litre vessel and the earlier 940 cc vessel was in the mechanism for stirring the reactor contents. In the smaller vessel, following modification to achieve continuous operation, a flip-flop stirring device with an external magnetic drive was employed. For the larger vessel W R D Manning, who was responsible for much of the design of high pressure equipment used in the scale up of the polyethylene process, and in the formulation of a theoretical basis for such design, proposed the use of a rotary stirrer. He further suggested that an induction drive motor should be incorporated alongside the reaction space and that the motor should be designed to operate in ethylene at reaction pressure. The motor would be cooled by incoming gas and the stirrer shaft would extend into the reaction space. A number of different variants of such motors and stirrers were evaluated during the work on the 9 litre and subsequent vessel designs, but the same principle is still employed in the most modern autoclaves working in the polyethylene process today.

Techno-Commercial Developments 1937/1938

In August 1937, following the electrical property work by Metropolitan Vickers, a meeting was held with the Post Office engineering department to consider ways in which Alketh could be used commercially. A wide range of properties was tabulated covering not only electrical properties, but also low temperature results, mechanical strengths, melt properties and water permeability, etc. It was agreed that possible applications could be:

1)   Use as an insulating or spacing material in high frequency equipment generally, eg coaxial cables.

2)   As a sheathing material substituting lead on telephone cables.

3)   Coil impregnation.

It was also commented that having regard to its permeability, power loss and general physical properties, Alketh also seemed to have promise as a submarine cable dielectric, replacing gutta percha derivatives. Interest was also expressed by British Insulated Callenders and Cables in material with the properties of Alketh.

Even at this early stage in the development of the product, ICI was advising interested parties that, if all went well with the work in progress, it anticipated being in position to produce Alketh the product at a rate of 250 tonnes per year by mid-1939 and with a selling price comparable to that of high quality polystyrene. It is also worthy of note that, up to the end of January 1938, the total expenditure on the Wallerscote 9 litre pilot plant amounted to only £3,816/10/9!

During 1938 much of the work on the 9 litre reaction vessel was concentrated on improving the mechanical reliability of the equipment and on producing samples for commercial evaluation. By May of that year the production rate had been raised to between 1½ and 2 kg/hr over short periods, with the product molecular weight being dependent upon conditions. Later, in October, a record production of 580 lbs of a range of molecular weights of product was achieved, and, in the first two weeks of continuous shift operation in November, a total of 530 lbs was made.

On the commercial side, discussions had been continuing with the Cable Makers Association (CMA), which represented some 90% of the cable industry, and separately with Submarine Cables Ltd (SCL) and the Telegraph Construction and Maintenance Company (TC&M), who were not members of the CMA. The latter two companies started testing samples of Alketh in mid 1938 and expressed such positive interest that, in August 1938, the ICI Alkali Division Board sought and obtained sanction for a 70 tonnes per year production plant, based upon a new reaction vessel with a capacity of 50 litres, to be sited at Wallerscote, and with a planned start up date of end 1939.

Although negotiations between ICI and the CMA were broken off in December 1938 because of the unwillingness of CMA member companies to enter into any firm forward commitments for the purchase of Alketh, the continued enthusiastic and bullish outlook for the product, expressed by both SCL and TC&M, resulted in ICI increasing the design capacity of the proposed production plant to 100 tonnes per year, and bringing its start up date forward to September 1939. Subsequently, because of the substantial interest expressed by other potential customers for the product, sanction was sought and obtained, in January 1939, for a second 100 tonnes per year production plant, at an estimated capital cost of £40,000, to be started up in early 1940.

Here it is of interest to note that in his submission to the ICI Research and Technical Committee in December 1938, P C Allen, then Alkali Division Research Director, and subsequently Chairman of ICI, suggested that, in the future, up to 1,600 tonnes of Alketh could be used in the manufacture of two transatlantic telephone cables. He also proposed that an offer should be made to SCL to supply them with up to 150 tonnes of Alketh by the end of 1940. In the financial case for the additional plant, it was forecast that at a selling price of five shillings per pound, and with an estimated manufacturing cost of two shillings per pound, an annual profit of £70,000 could be achieved. This assumed that the actual plant production could be raised to 250 tonnes per year and made due allowances for capital expenditure.

In contrast to these optimistic projections were the actual sales of Alketh samples achieved over the 12 month period from December 1937. A total of 280 pounds had been sold to Du Pont in the USA, 117 lbs to SCL, five pounds to Standard Telephones and Cables, five pounds to the Royal Air Force in the form of moulded discs and 1 lb plus two samples of film to British Insulated Cables. All samples were invoiced at 15/- per lb, and the total income received by Alkali Division from its sales of 408 lbs of Alketh in 1938 was the princely sum of £306.

In March 1939 a new price structure was established for Alketh in order to prepare for the forthcoming production operation. This had the effect of more than halving the previous figure of 15/- per lb for all but the smallest customers, with the structure being:

	2 cwts or more	5/- per lb
	1 cwt	6/- per lb
	56 lbs	6/6 per lb
	28 lbs	7/6 per lb
less than	28 lbs	9/- per lb

Between the announcement of this new price structure, and the outbreak of war in September 1939, 100 contacts were established with potential customers for Alketh, and both SCL and TC&M developed promising outlets for the product in high frequency cable fields, and sold some lengths of Alketh coated cable for television work. Over the same period some 2,600 lbs of Alketh was sold to customers, most of which went to the cable manufacturers.

The 50 Litre Reaction Vessel Plant

It will be recalled that the 9 litre pilot plant started operation in November 1937, but continuous operation was not achieved until late in 1938. During this period the operation consisted of feeding gas to the reactor, at a pressure of up to 1,500 atmospheres, which had been compressed by the continuous mechanically driven mercury compressor, developed by Michels. From this reactor the products passed through a specially designed let down valve into a hopper held at near atmospheric pressure. From this, ethylene was blown off to atmosphere and the liquid polythene was blown down into a container where it was allowed to solidify, and subsequently granulated.

In December 1938, a major step forward in the process was made when a new design of hopper and extrusion valve was introduced, which enabled continuous operation to be achieved. In this, ethylene was recycled from the hopper for recompression and reuse in the process, whilst the molten polymer was blown down via an extrusion valve into a container for solidification and subsequent granulation. The actual design of the first 50 litre reaction vessel had been completed by Manning in 1938, and orders were placed in December of that year for two complete vessels plus billets for a further four.

Although all high pressure compression experience had previously been gained with mercury compressors of the Michels design, an order had been placed with Peter Brotherhood in March 1938 for a water lubricated piston compressor for compression of ethylene from 250 atmospheres to 1,500 atmospheres, at a rate of 96 lb/hr and at an estimated cost of £760. However, because of the experimental nature of this machine, a hyper mercury compressor, using the proven Michels principle, was designed by ICI engineers for delivery of up to 130 lb/hr of ethylene at a pressure of 1,500 atmospheres, for incorporation in the new plant.

The basis of design of this plant was thus a primary piston compressor delivering gas at 250 atmospheres to a mercury hyper compressor which fed the gas to two 50 litre reaction vessels operating in parallel. The products of reaction from both vessels were discharged via specially designed let down valves into a single hopper system, from which unreacted gas was returned to the process for recompression, whilst the molten polythene was blown down into moulds to solidify. Subsequently, blocks of the polymer weighing about 16 lbs and measuring 24 by 10 by 2 ins, were cut into slices by a machine similar in design to an agricultural turnip cutter.

Fig.10. 50 Litre reaction vessel installed in bay. Courtesy of ICI.

Since orders for the equipment had been placed shortly after achieving continuous operation on the 9 litre pilot plant, a conservative basis had been used for the sizing of the various vessels and machines for the production plant. However, when the 130 lb/hr hyper compressor was delivered, it was possible to evaluate the effect of feeding gas at higher rates to the 9 litre vessel. These experiments showed that the yield of polymer per unit volume could be increased as the gas rate increased but not in a directly proportional ratio. Thus a four fold increase in gas rate only gave a 2½ times increase in polymer production. However, as a result of this work, it was clear that two 50 litre reaction vessels could be expected to produce up to 200 tonnes per year of polymer, provided that a total of 900 tonnes per year of compressed gas could be provided.

In the event, the first 100 tonnes per year production plant was completed with two reactors operating in parallel in June 1939, and the first polyethylene was produced in September of that year. Later that month, continuous operation was achieved and between then and the end of the year production rates rose steadily to an equivalent of 36 tonnes per year. At the same time work was continuing on installing equipment for the second 100 tonnes per year plant but, because of supply problems, this was not completed until May 1940.

In the first half of 1941, when the two units were in operation, a total production rate equivalent to 165 tonnes per year was achieved, still using two reactors working in parallel on each stream, and using the experimental piston compressor only when breakdowns occurred on the mercury hypers. Later that year a presentation was made to the ICI Alkali Division Board of the various ways in which output could be increased from the two streams, up to a total of 328 tonnes per year. Following this presentation, and as a result of the substantial increase in demand to meet wartime market needs, sanction was obtained to install a further 250 tonnes per year plant at Wallerscote, which was completed by the end of 1942.

By 1946, the Wallerscote plant consisted of 5 reaction bays, each containing its own 50 litre vessel, three primary piston compressors capable of delivering a total of 1,300 lb/hr of ethylene at 250 atmospheres, four mercury hyper compressors with a combined pumping capacity of 505 lb/hr and four piston secondary compressors with a combined capacity of 600 lb/hr. Up to 300 lb/hr of ethylene was available from the ethylene plant and total polymer production amounted to between 700 and 800 tonnes per year.

The 250 Litre Reaction Vessel Plant

Following the outbreak of war in September 1939, and as a result of the forecast requirements for polyethylene for use in wartime applications (dealt with below), ICI started the design of an even larger reaction vessel with a capacity of 250 litres. By December of that year a design for a plant to produce 250 tonnes per year had been completed, with an estimated capital cost of £50,000. An economic appraisal suggested that a two stream plant based on the 250 litre vessel, together with the 200 tonnes per year Wallerscote plant, could produce polyethylene at a cost of less than 2/- per lb. It was further estimated that, post war, the product could be sold at 3/6 per lb to compete with gutta percha, thus yielding an annual profit of over £87,000.

Fig.11. Reaction bay after ductile failure of 50 litre vessel and explosion. Courtesy of ICI.

In February 1940 the ICI Board gave approval for the construction of a 500 tonnes per year plant on the basis of the interest shown in the product by TC&M, Du Pont and Les Cables de Lyons in France. Subsequently, discussions were held with the Ministry of Aircraft Production, who gave an A1 priority to the building of this plant, but stipulated that it must be sited at least one mile distant from the Wallerscote plant. A new site was therefore selected at Winnington, and the plant construction was completed by August 1941 at a total cost of £200,000, ie over twice the original estimate! The first polymer was made on the plant in November 1941, but in the following January a major explosion occurred, following an ethylene leak.

Fig.12. 50 litre reaction vessel after ductile failure. Courtesy of ICI.

Figures 11 and 12 show the devastation which can occur following such an explosion. These relate, however, to a 50 litre vessel explosion in which the relief vents became blocked. Such was the demand for the product however, that permission was given for blackout restrictions to be ignored, to enable repair work to be continued round the clock. By March, rates of between 100 and 300 lbs/day were established, but a succession of leaks, fires and explosions continued to prevent steady operation. However, by June the make rate had been increased to more than 1,000 lb/day, although varying between 26 and 104 lb/hr, and with melt flow indices (a function of molecular weight, the lower the index, the higher the molecular weight), ranging between 2 and 93. By December 1942 the make rate had been increased to 2,000 lb/day, with hourly rates ranging between 17 and 97.5 lbs, and by late 1943 to 2,700 lb/day, with hourly rates being much more consistent, between 89 and 124 lbs. By this time closer melt flow index control was also being achieved, with a range between 11 and 32. Figure 13 shows the control panel for the reaction vessel and the vessel operator controlling the vessel pressure by means of a driving wheel connected to the reactor let down valve.

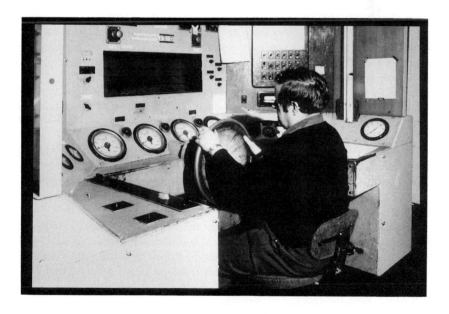

Fig.13. Pressure control of reaction vessel using driving wheel. Courtesy of ICI.

In September 1941, a review was carried out of the potential for increasing the capacity of the Winnington plant, and in 1942 approval was given for the installation of a third 250 tonnes per year unit. For this expansion, Government "Grant in Aid" was sought and obtained for a £40,000 loan to be repaid over a period of 6 years, with an interest rate of 6%. In the event, ICI did not take up the loan, and also proceeded to sanction a fourth unit for Winnington in January 1943. Start up of the third unit was achieved in early 1944 and of the fourth in

October 1945, delays with both having occurred because of problems with the allocation of steels, which resulted in representations being made by the ICI Chairman direct to Government ministers, before scarce stocks were allocated to the polyethylene extension.

By the time the fourth unit started up at Winnington, the plant consisted of 3 Weir primary compressors with a combined capacity of 1,500 lb/hr, 4 mercury hyper compressors with a total capacity of 1,800 lb/hr and 4 x 250 litre reaction vessels. The ethylene plant had a peak capacity of 1,100 tonnes per year. It is also of interest that during the war years many of the plant shift operators were women and it was not until December 1945 that they were replaced by men!

The Market - 1939 to 1946

At the time of the outbreak of the Second World War in September 1939, the bulk of the production was being used in high frequency cable applications, with some usage in wax candle manufacture. By the end of 1939, TC&M and SCL had taken 12,856 lbs, Hydroplastics (another cable manufacturer) 4,652 lbs, and Prices Candles 1,796 lbs.

At the beginning of 1940, TC&M placed orders for the delivery of 159 tonnes of polyethylene, 23 tonnes of which was required for high frequency cables (duty 10,000 volts at 200,000 cycles per second) and 136 tonnes for submarine cables. A further 60 tonnes was estimated as being required by the Air and War Ministries and 200 tonnes was the estimated requirement by the Royal Engineers and Signals Board for twin and quad cable insulation. Other forecast offtakes were 50 tonnes by Hydroplastics for HF wire covering and also for the incorporation of luminous powders, 5 tonnes by Prices Candles and 20 tonnes by Les Cables de Lyons. In the event, only 105 tonnes of product was manufactured in 1940, of which 95.8 tonnes went to Telcon companies (TC&M and Hydroplastics).

In mid-1940, four-ninths of production was allocated to Ministry of Aircraft Production outlets with the remainder being equally divided between the Ministry of Supply and Admiralty requirements. Later the Admiralty allocation was reduced since its use of polyethylene was considered to be only convenient, rather than essential!

In 1941, Callender and Cables joined the Telcon companies in producing Quad cable for the Ministry of Supply and 75% of the total production of 176.91 tonnes in that year was sold to these companies: British Insulated Cables took 14 tonnes, and Pirelli General Cable Co, Tenplas Wire and Sleeving Co, Metropolitan Vickers, Prices Candles and Western Electric in the USA took the bulk of the remainder. With the collapse of resistance in France and the consequent demise of the requirement for candles in the trenches, Prices Candles were removed from the list of approved companies for allocation of the product in mid-1941, and virtually all of the production was subsequently allocated to high frequency cable uses.

It is of interest to note that in none of the records of the period was any mention made of RADAR, to which the bulk of the production was allocated. The secrecy surrounding this invention was maintained until late in the war and, as a consequence, there were virtually no scientific papers published covering developments in polyethylene manufacture or applications. With the outbreak of the war with Japan in December 1941, the Post Office's interest in replacing natural rubber by polyethylene increased sharply. However, because of the inability of the Wallerscote and Winnington plants to meet demand, a rationing scheme was instituted between December 1941 and April 1942. Actual forecasts for the market requirements in 1942 amounted to more than 700 tonnes, but in the event only 557 tonnes were manufactured, of which Telcon took almost 50%.

During 1941, as a result of forecasts that production would be unable to keep up with demand, work was carried out in the newly created Plastics Division of ICI to determine the effects of blending polyisobutylene (PIB) with polyethylene. It was found that up to 12½% of the rubber could be incorporated without a deterioration in electrical properties, but in January 1942 only 10 tonnes of PIB was available in the country, the only available source being the USA, where the total manufacturing capacity was only 45 tonnes/month. However, negotiations involving the Ministry of Supply resulted in sufficient supplies being made available to meet essential requirements over the production shortfall period in 1942, and PIB was subsequently used in all cable product to reduce the risk of cracking.

In May 1942 ICI introduced the trade mark "Alkathene" to replace the old name of Alketh. Later that year Metropolitan Vickers developed an extrusion moulding process for making waterproof cable joints and round dipole aerials etc. Rods, bars and solid blocks were also produced for the manufacture of switches for high frequency electrical equipment.

Early in 1943 the selling price of the product was reduced to 4/- per lb for 80 lb lots and the product range consisted of melt indices 2, 7, 20 and 70, plus unfiltered material. Of this range, only TC&M were willing to take up to 4 tonnes/month of MFI 7, the remainder of the cable requirements for all manufacturers being met by MFI 20 product, whilst MFI 70 was supplied to Metropolitan Vickers alone.

During 1943, some of the manufacturing bugs in the process were resolved and, with extra capacity coming on stream, a total production of 1,049 tonnes was achieved, with only 920 tonnes being sold to customers. Of this, TC&M took 41%, Callender and Cables 16.5%, and 23.5% was delivered to ICI Plastics for blending and onward sale. During this period film 0.00125 ins thick was made and ICI Plastics worked on a process for film manufacture which they considered could make film of less than 0.003 ins thickness for sale at 10/- per lb, compared with an estimated conversion cost of 2/- per lb.

As a result of the growth in stocks of certain grades (especially of MFI 2 product, which was incapable of being extruded on available equipment, of MFI 7 of which 37 tonnes was in stock by mid-1943, and MFI 70), new applications were sought in submarine cable, high tension power cable, mining cable, low tension aircraft cable fields etc. At the same time the first sales pamphlet on the ICI Alkathene brand of Polythene was published.

Another development in 1943 was the addition of antioxidants to the polymer to prevent deterioration of power factor during processing.

In 1944, when the third Winnington unit started up, total production rose to 1,441 tonnes, of which TC&M took 317 tonnes, Callenders 317 tonnes, SCL 162 tonnes, and Pirelli 138 tonnes, with 351 tonnes being processed before onward sale by ICI Plastics. Also in 1944, a number of Alkathene insulated telephone cables, for use at radio frequencies, were laid across the Channel to support the Normandy invasion.

In 1945 the Alkathene price was reduced to 3/3 per lb for standard lots, and a range of coloured compounds was introduced. However, because of the expectation that the war would soon be coming to an end, the offtake of Alkathene by the cable producers dropped substantially during 1945, with TC&M reducing their offtake to 143 tonnes. SCL were more bullish about prospects for the product and were planning a 13 mile submarine cable between the Isle of Wight and the Mainland, which was eventually laid in 1946 and used 2½ tonnes of Alkathene per mile. They were also planning 3 inch, and possibly 8 inch diameter cables between the UK and Holland, Norway and Poland, and were forecasting a consumption of between 1,300 and 5,000 tonnes of Alkathene for this application within 5 years of the end of the war.

As well as the use of the product for submarine telegraph cables, low voltage airborne cables, power cables etc in 1945, 25 tonnes of Alkathene was used in the form of an MFI 70 compound, incorporating PIB, for film to provide a moisture-resistant packaging for Mepacrine tablets.

In spite of these new applications, the fall off in demand following the end of the European war in May, and of the Pacific war in August of 1945, resulted in an unacceptable increase in stocks. Two units on the Winnington plant, together with all units on the Wallerscote plant, were consequently closed down in November, and the output of the third Winnington reactor limited to 2,000 lb/day. However, by April 1946 all three units were back in operation and by the end of November of that year were operating at a rate equivalent to 1,380 tonnes per year.

In the search for new applications following VE day, callendered polyethylene sheet for lampshade applications was introduced under the trade name "Crinothene". A very low molecular weight polyethylene was made for wax applications, by a hydrogen modification process, with the trade name "Winothene", and a chlorinated polyethylene was produced with the name "Halothene". None of these was to prove the saviour of the polyethylene business and it was not until 1947, when ICI Plastics started its first experimental 48 inch-wide film extrusion unit, following the lead taken in the USA, that the future appeared to brighten for the product. That, however, is another story!

Other Producers - 1939 to 1945

As has been mentioned previously, under technical collaboration arrangements with Du Pont in the USA, samples of polyethylene had been provided to them as early as 1936. They continued to evaluate the product during subsequent years and introduced the material to Western Electric, who ordered 2,000 lbs in 1940 for co-axial cable work. At the same time, in the early 1940s, they started work on the design of their own production plant based upon a tubular reactor process.

In September 1942, the ICI Chairman, Lord McGowan, contacted the Du Pont President and offered to show them the ICI manufacturing process. In October, three Du Pont engineers arrived in the UK to obtain details of the ICI process which might be applied to their own two stream 670 tonnes per year tubular reactor plant. A royalty agreement was drawn up, but it was agreed that Du Pont would be excused payment during the wartime period.

The Du Pont plant, for which the bulk of the equipment was reported to have been paid for by the US Government, started up in May 1943, and during the period until December of that year manufactured 195 tonnes. The plant was then shut down for modification because of severe manufacturing difficulties encountered, which were largely associated with tube blockages.

In 1944 production was resumed using a wet tube process, in which a mixture of ethylene, benzene and water was fed into the tubular reactor. By using this technique, it was hoped that heat transfer would be increased through the tube walls by eliminating polymer build up, and also to increase the heat sink for taking up the heat of polymerisation produced in the highly exothermic reaction. This process was continued until after the end of the war in reactors about 300 metres long and with an internal tube diameter of 2.8 cm. Sales of polyethylene by Du Pont, including imports from ICI, were 236 tonnes in 1943 at $1 per lb, 298 tonnes in 1944 at $0.83 per lb and 707 tonnes in 1945 at $0.73 per lb.

In April 1943 Du Pont had been granted rights to grant non exclusive, non assignable sub licences for user purposes, under ICI's US patents. They were also advised that they could offer a manufacturing sub licence under ICI patents to Carbide and Carbon (now Union Carbide Corporation), who had also developed a tubular reactor process for the product which was based on information contained in the ICI patents. In the event, Carbide and Carbon decided that such a licence was unnecessary, and did not conclude such an agreement until after the end of the war. They did, however, achieve substantial success with their dry tube process, in which ethylene at a pressure of about 2,000 atmospheres pressure was fed, together with a small percentage of a high-boiling hydrocarbon incorporated to prevent incipient tube blockages, into a reactor believed to be some 55 metres long with a tube ID of 0.8 cm. By the end of the war a production rate of around 1,100 tonnes per year (a figure substantially in excess of that which had been agreed with the US Navy, because their units had been found to be "much more productive"), was being achieved from each reactor. By 1946 Carbon and Carbide were reported to have achieved sales in the USA for their product at a rate of 3,243 tonnes per year.

The only other producer during the wartime period was IG Farben in Germany, who developed a 260 ft by ½ inch diameter dry tubular reactor, again based upon ICI patent information. This reactor worked at a pressure of 1,100 atmospheres, and had a production capacity of 5 to 10 tonnes/month of product with a molecular weight of between 15,000 and 20,000, and up to 20 tonnes/month of a product with an molecular weight of 2,000 to 3,000.

In 1942 a team from the USSR visited ICI under the terms of a secret Government to Government treaty for the exchange of technical information. They were offered technical information, free of charge, for a manufacturing plant for polyethylene, with the proviso that after the cessation of hostilities they could manufacture up to 2,500 tonnes per year for a once off payment of £250,000. Nothing further was heard about their progress, although some unsuccessful attempts are believed to have been made to develop a production operation.

Postscript

The end of the war basically saw the end of the first phase of the polyethylene story. If the film market had not developed in the late 1940s, it is doubtful if ICI would have proceeded to license its technology to other European producers in the early 1950s, although the Ryan judgement in the USA would still have forced the licensing of the technology to other USA producers.

In the event however, ICI decided to license its technology and, together with its engineering contractor Simon Carves Ltd, has been responsible for more than 30 plants, incorporating 70 reactor streams, for the manufacture of high pressure low density polyethylene operated by licencees worldwide. Approximately 50% of the total world capacity for the product is still produced today on plants using autoclave processes derived directly from those described in this paper, with the largest ICI technology reactor streams having capacities in excess of 100,000 tonnes per year for a range of different grades.

Acknowledgements

The author would like to acknowledge the assistance given by ICI Chemicals & Polymers Ltd, in making available reports and documents relating to the development of the ICI polyethylene process during the period 1935 to 1947, and for permission to publish this paper. Thanks are also due to staff at the Science Museum Library, who made available documents, again relating to the process, which had been lodged there by ICI.

Plastics and Prosperity, 1945–1970

Plastics: The New Engineering Material

Martin Thatcher

GREAT LAKES CHEMICALS EUROPE LTD, ELLESMERE PORT,
CHESHIRE, UK

Introduction

This paper will concentrate on a group of plastic materials widely known as "engineering thermoplastics". Since this is a collection of materials whose commercial history commenced very close to 1945, it is interesting to see how the rapid progression of these materials relates to the many factors which have influenced industrial and consumer activity from the end of World War II to the present day.

This period of "prosperity" is characterised by continued growth of mass consumption and mass production, created following the strictures of the war. This favoured the investment in "New Engineering Materials" which could offer efficiency in the manufacture of consumer items. These consumer items benefitted from the materials since, in addition to offering the right level of performance and durability, they were light-weight, colourful and could be contoured both for ease of use and product differentiation. The combination has resulted in the sales of this group of raw materials soaring from a few hundred tonnes worldwide in 1945 to some 3 million tonnes per annum today, and growth rates averaging 12% since 1970.

Engineering is a part of all stages of the production process: from the ability of the chemist to engineer molecular structures most likely to give desirable properties; through the engineering of industrial scale plant capable of producing large quantities of high quality polymers and compounds; to the capability of the end-product developer to understand the properties of these new materials and apply them in design; and finally, advances in production engineering particularly injection moulding technology. Progress in all these elements has been achieved and applied at a far greater pace than has happened in the past with any new class of materials. This paper will trace some of the driving forces behind this phenomenal development.

Towards a definition of engineering thermoplastics

The term engineering thermoplastics is itself the brain child of a unrecorded post-war marketeer and consequently is a somewhat "elastic" concept generally applied to rigid thermoplastic materials with a degree of mechanical integrity at temperatures around 100°C. The author has been unable to trace when the term ' was first used, although E.I. du Pont de Nemours (Du Pont) literature in the 1950s refers to a class of engineering materials, and Celanese Corporation publicity refers directly to engineering plastics in the launch campaign for its acetal copolymer in 1962. Whilst plastics were widely known to the public, the chemical companies involved in the development of new materials sought to differentiate a superior class of plastic material which could be seen as a substitute for metals in demanding application areas. Doubtlessly spurred on by the success in establishing "Nylon" as a new class of textile fibre, the challenge was to place the moulding material variant equally firmly as the material in the mind of the engineers and designers looking to build and shape the post war consumer products.

The engineering thermoplastics encompass a broad range of chemistry in arriving at the final commercial grades which number many thousands. Initially, of course, there are the polymers which through their composition and structure define the basic properties that can potentially be achieved. To enhance the properties in use and maximise the retention of properties over time an array of additives are employed, which are, in turn, the product of the chemical industry. These include antioxidants, · light stabilisers, metal de-activators, plasticisers, lubricants, colorants, and flame retardants. The use of reinforcing fillers to increase rigidity and short term thermal performance is now widespread, especially in the semi-crystalline materials where the reinforced variants often represent the bulk of commercial grades and sales volume. As the engineering thermoplastics materials have developed and the tailoring of grades of material towards a set of particular end-user application requirements has increased, so the mechanical blending of two or more polymers has become an ever more important aspect of product development in achieving the desired property profile.

The engineering thermoplastics give the designer a range of materials with relatively predictable performance characteristics to which standard engineering equations can be applied. The general properties influencing a material's selection include strength, stiffness, creep, fatigue, toughness, flammability, chemical resistance, electrical properties, with most or all of these properties being maintained over an extended range of temperatures. In addition to these characteristics, the rapid adoption of these materials has been driven by the cost benefits compared with traditional materials. The ability to injection mould products to tight dimensional tolerances in very high volumes matched the need to produce large numbers of consumer goods at low prices which fuelled the post war expansion of the economies of the western world.

The history of engineering thermoplastics as a reflection of chemistry and society in the twentieth century

All chemists involved in the development of polymer systems owe a debt to the empirical work on organic chemistry carried out in the latter part of the nineteenth century. Chemists at that time observed the formation of viscous fluids during experimental work on a variety of different organic reactions. They were regarded as interesting but did not result in further investigation of thermoplastic systems.

A major watershed was reached in the 1920s when Hermann Staudinger, Hans Meyer and Hermann Mark proposed the structure of macromolecules as valence bonded, high molecular weight entities and not as special aggregates of small molecules.

In 1927 Dr C.M.A. Stine, Director of the Du Pont Chemical Department, urged his executive committee to support a request to fund fundamental research on organic chemistry, which led to the hiring of Dr Wallace Carothers. The subsequent work of his team on both the theory of polycondensation, and the development of polyesters and polyamides is well known. Carothers' efforts were focussed on the need to produce synthetic materials as textile fibres, with the desirable properties of strength, chemical resistance and heat capability related to the ability to produce woven textiles that could be cleaned and pressed by conventional means, without any thoughts of their use in solid moulded components. Such rapid progress was made following the laboratory work to produce a "superpolymer" from hexamethylenediamine and adipic acid in February 1935, that the first polyamide 66 fibre plant went into full production in early 1940. A polyamide moulding powder was made commercially available in 1941, with its first application being for electrical coil formers. In this same year I.G. Farben in Germany launched polyamide 6 as a moulding material. A year later these products together with polytetrafluoroethylene (PTFE) which had also come out of the laboratories of Du Pont were classified as strategic materials and all production was dedicated to the war effort.

In 1945 whilst the use of the materials during World War II had revealed their versatility, it was evident that a level of technical support would be necessary if they were to be promoted and used effectively in demanding applications. Hence the establishment of a Technical Service Laboratory by Du Pont in Arlington, New Jersey, to promote the use of nylon (or ZYTEL as the moulding material was trade named in 1950) to replace metals. This technical marketing effort has since gone from strength to strength during the intervening decades.

The energies of the chemists and the financial resources of their masters were now unleashed in the search for other thermoplastics with desirable property profiles. It is not surprising that much of this effort was focused in the United States. The table below provides a rapid review of the resulting flurry of patents and publications in the 1950s:

1948	Pure formaldehyde monomer prepared
	Styrene-Acrylonitrile/Buna N and S blends patent
	Polyvinylidene fluoride patent
1950	Polyamide (PA) 6 continuous caprolactam process
	PTFE commercial plant, Du Pont
	I.G. Farben broken up
	PA6 ULTRAMID - BASF trademark
	PA6,6 ZYTEL - Du Pont trademark
1952	Stable end-capped polyformaldehyde - Du Pont

1953	Polycarbonate synthesised - General Electric, USA
	Polycarbonate synthesised - Bayer
	Styrene-acrylonitrile-butadiene emulsion polymerisation - Marbon Division, Borg Warner Corporation
1954	ABS commercial - Marbon Division, Borg Warner
1955	PA11 commercial plant - Thann and Mulhouse, France
1956	Polyacetal "pre-commercial" field trials
1959	Polycarbonate commercial LEXAN/ MAKROLON
	Polyacetal commercial DELRIN

This listing only scratches the surface of what was going on in terms of intermediates, catalysis, and stabilisation, as well as chemical engineering and the fundamental understanding of polymer structures and reactions.

The 1950s saw significant work on two engineering thermoplastics products which do not derive from textile fibre research: the high impact, transparent polycarbonate amorphous products from General Electric, USA, and Bayer AG, Germany, and the polyacetal homopolymer, a semi-crystalline material from Du Pont and the related copolymer from Celanese Corporation.

In terms of engineering and applications this decade saw a great step forward when Ankerwerk of Nürnberg, Germany introduced the first reciprocating screw injection moulding machine in 1956. This had its origins in work done by H. Beck in Ludwigshafen in the 1940s. The control of the heating and plastification was far better than in the previous plunger type machines giving a homogeneous melt and facilitating the consistent production of accurate components.

In "marketing" terms the decade produced some notable firsts. The work of General Electric (GE) has already been mentioned but it must be seen as having particular significance as this was an "end-user" entering the fray and not a traditional chemical company. GE was a huge user of electrical insulating materials with resources to allow its corporate research staff to pursue avenues which would demand very considerable investment to put into effect. That it was taking the first steps in "strategic planning" which were to become a hallmark of American corporate management practice, reinforced the significance of its involvement in plastics. It can be argued, with some justification, that the commercial manufacture by GE of polycarbonate relied very heavily on industrial scale processes licensed from Bayer AG. Nonetheless the end use perspective is one that has dominated the GE approach to building a portfolio of products, based solely on engineering thermoplastics, with some speciality additives, which accounted for sales in excess of 4.5 billion dollars in the financial year 1991, and to the claim of world leadership in this sector.

In 1954, Du Pont expanded its service capability to the market when it brought into operation a new, more customer oriented Plastics Technical Service Laboratory in Wilmington, Delaware. In 1958, the processing laboratory was equipped with a reciprocating screw machine which was used as the basis of the processing work on Du Pont's polyacetal (polyformaldehyde) material. Du Pont's experience with the introduction of polyamide as an engineering thermoplastic material convinced it that careful preparation was needed, both through in-house and field experience, prior to the full commercial launch of the DELRIN product.

The "swinging sixties" saw the continuation of new product introductions, as tabulated below:

1962	Acetal copolymers - Celanese Corporation, USA
	PC/ABS blend patents - Borg Warner, USA
	Polyphenylenesulphide linear polymer - Dow Chemical, USA
1965	Polysulphone - Union Carbide, USA
1968	Polybutyleneterephthalate - Celanese
	Polyphenyleneether/polystyrene blends - General Electric

This decade also saw Akzo of The Netherlands introduce injection moulding grades of polyethylene terephthalate, and melt processible fluorocarbon polymers from Pennwalt and Du Pont. Much work continued on copolymer systems and an increasing array of property enhancing additives. Foremost among such additives was the use of chopped strand, glass fibres as a reinforcing medium. Considering how widespread glass reinforced engineering thermoplastics are now, it seems hard to credit that they have only been available for twenty five years or so. Flame retardant additive systems also started to be applied at the end of the decade.

The introduction of NORYL in 1968 by General Electric also sheds light on its end-user perspective. Both GE and Akzo worked on polyphenylene ether systems, however this high temperature amorphous material degraded severely at the required processing temperature. By chance, it was discovered that the polymer was completely miscible with polystyrene (PS). Thus, by blending these two materials in a melt extruder, compositions could be formulated with glass transition temperatures between that of PPE (208°C) and PS (100°C). Thus short term heat capability could be targeted to meet key customer requirements and specifications. In addition the PPE/PS blend required lower processing temperatures, and phosphate esters proved an effective flame retardant/plasticiser. Hence, the first significant blend product in engineering thermoplastics was aimed at niche markets lying between the capabilities of the styrenics, including ABS, and the polycarbonates.

The 1970s and beyond saw the high performance materials with long term temperature capabilities above 200°C materialize, with commercialisation of the sulphur containing polymers:

> Polyphenylenesulphide (PPS) - Phillips
> Polyarylethersulphone (PES) - ICI
> Polyphenylsulphone - Union Carbide
> Polyarylethersulphone - Union Carbide

and two non-sulphur systems:

> Polyether ether ketone (PEEK) - ICI
> Polyetherimide (PEI) - General Electric

Further work on blends saw the introduction in the 1970s of two significant products, namely polycarbonate/ABS from Bayer AG aimed at the flourishing PPE/PS market developed by GE, and polycarbonate/polyester blends developed by GE Plastics European division, with a specific application in automotive bumpers, for Ford Europe.

Polyacetal: a case history

This material has its origins in Butlerov's description of formaldehyde polymers in 1859, but only became commercially feasible following the availability of formaldehyde monomer of high purity in 1950. This was followed by the work of R.N. Macdonald, who was also influenced by the introduction of polyamide as a moulding material by Du Pont. Thus it was in 1952 that scientists at Du Pont achieved a continuous polymerisation using a polymeric tertiary amine initiator in a liquid hydrocarbon medium to produce a high molecular weight polyformaldehyde. The pure acetal homopolymer was prone to "unzipping", but stabilization was achieved by "end-capping" in solution with acetic anhydride.

A remarkably rapid decision was reached by Du Pont management to commercialise the product as a thermoplastic moulding resin. Thus it became the first hydrocarbon engineering thermoplastic to be developed outside the textile fibre field. It also benefitted from the experience being gained with marketing the polyamides. Under the code name "Polymer F" the work continued to develop a viable product and process, given the inherent ease with which the polyformaldehyde material can degrade back to the monomer and the rather rudimentary temperature control available on the plunger type injection moulding machines, the development was an act of considerable faith. Five years later, in 1957, the board of Du Pont authorised the building of a commercial production plant. At this point, three years prior to commercial manufacture, the marketing effort began in earnest to define what key properties should be specified in order to meet the needs of the customers to produce accurate and consistent components. This is an early example of industrial marketing on this scale. In addition to defining the product specification, characterisation studies and field trials continued over the period.

The result was that, at its launch on 1st January 1960, the product under the DELRIN trademark was supported by unparalleled technical literature aimed to help the design engineer to make the switch from metal to plastic, and, of course, by a dedicated team of experienced technical specialists. This was a time of growing international horizons for industry and the DELRIN product was marketed through Du Pont International SA in Geneva to the European markets. The combination of a strong trademark, with full technical support aimed at both specifier and processor, and the international scope to the product introduction broke new ground as an approach to the marketplace, and became a benchmark for others to follow.

Although Du Pont's patent lawyers had attempted to protect its technology as best they could, they had left one chink in its armour: the Du Pont patent covered only the homopolymer. Given the strong competition developing among key players to take a share in this new breed of materials, someone was bound to exploit the weakness. Chemists at Celanese Corporation produced a copolymer material in 1957, which was produced by the cationically initiated polymerization of trioxane in the presence of ethylene oxide or 1,3-dioxolane. Significantly, a greater degree of stability to thermal oxidation and alkaline attack can be achieved with the copolymer than with the homopolymer. The copolymer is subjected to a thermal treatment which "unzips" the polymer back until a carbon-carbon bond is reached. This is done in the presence of antioxidants to prevent random oxidative attack of the polymer chain to give a very stable end-group. In 1962 Celanese introduced their copolymer under the CELCON trademark, by which time Du Pont already had 2250 tonnes of sales under its belts. Both companies had two grades, one moulding and one extrusion, with typically the second player, Celanese, positioning their products with slightly lower molecular weights and therefore better flow. The response was also typical when in 1963 Du Pont introduced an even lower molecular weight grade.

In the USA, the introductory truckload price of $0.95/lb on 1 January 1960 was dropped to $0.80/lb by 1 June that year, perhaps in an attempt to frighten off the potential competition. The interesting price evolution of acetal over the years can be traced using data published in the October 1992 issue of Plastics News by K.J. Persak & R.A. Fleming of Du Pont:

Year	Price $/lb
1960	0.95
1970	0.65
1980	1.04
1986	1.64
1992	1.25

Considering the ravages of inflation over the period, the price has remained remarkably stable.

The international marketing effort continued with Du Pont starting the compounding of product in Europe in 1963. Celanese decided on a different route to the international markets initially joining up with ICI for UK sales of the US product. Ultimately Celanese licensed Hoechst AG to produce the copolymer. Ironically Hoechst AG subsequently purchased the Celanese Corporation in 1989.

In terms of grades, other than viscosity variants, 1964 saw PTFE filled versions of the homopolymer and a milled glass fibre reinforced version in 1965. Early glass filled grades from Celanese, whilst having improved flexural strength, showed poor tensile properties. This was remedied by the development of a glass coupling system. By the 1970s, injection moulding technology had advanced to the point where polyacetals were being moulded in thin wall parts, on multicavity hot runner tooling, notably in the "Bic" disposable lighter which was at one time the largest world-wide consumer of this engineering resin. The 70s therefore became a decade of enhanced productivity grades for better flow and low mould deposit from additives and volatiles. The decade is also notable for the entry of Asahi Chemical Industries as a supplier of homopolymer, and Mitsubishi Gas Chemical who joined Daicel to produce copolymer in Japan.

The 1980s saw the so-called "super tough" elastomer-modified grades emerge, as well as the disappearance of Celanese Corporation. In a significant move, in 1990, Du Pont introduced DELRIN II, a copolymer, which in the eyes of some finally admitted that you could not do everything with a homopolymer. The competition for greater productivity from the materials continues today and will do so for many years to come.

Conclusion

These "New Engineering Materials" fifty years on are now an established part of the engineer's range of available materials. Have they deserved the "engineering" epithet that was thrust upon them by an eager marketeer anxious to differentiate them as having a technical pedigree and demanding a higher price? Certainly the development of the engineering thermoplastics has responded to the demands of the engineer:

value engineering	- for metal replacement
design engineering/	
computer aided engineering	- optimise cost/performance in use
production engineering	- from material handling to finishing

As products of the market driven era, the engineering polymers have been subject to numerous modifications in order to adapt to market needs. A select but highly motivated set of competitors vied for leadership in the increasingly global marketplace for engineering thermoplastics.

While engineering is still a part of the equation, and marketeers continue to invent "new" engineering concepts to support the value added sell, the novelty of the products is wearing rather thin. Increased competition and a harsh economic climate have eroded margins, improved commodity materials have made major inroads and producers are seeing diminishing returns from increasingly complex and expensive marketing programmes needed to break into totally new application areas. Even when these new applications can be secured, the honeymoon period of high margin and little competition is often very short lived.

So a degree of maturity is setting in, but this does not mean that all the excitement has gone out of this sector of chemical research and development. Far from it! Engineering thermoplastics are the product of innovation in the chemistry of materials and have carried that legacy of innovation through the initial fifty years of commercial development.

Versatility of Acrylics, 1934–1980

J. P. Tilley

ICI ACRYLICS, PO BOX 34, DARWEN, LANCASHIRE BB3 1QB, UK

Acrylic is a generic term for a very wide family of chemicals that can be converted into a variety of polymer forms, that is fibres, resins, powders, highly filled liquids and sheets. These are used for a wide range of applications. From carpets to clothing, from signs to kitchen sinks, from baths to car body parts and even paints the list is endless. I cannot hope to cover all of them in this paper so I am going to concentrate on one, namely "Perspex" acrylic sheet.

Where does it all start?

In the late 1920s and early 1930s considerable research work was being done in England and Germany into the basic techniques of producing polymethyl methacrylate. Within ICI, Dr Rowland Hill of the then ICI Dyestuffs Division was the first to notice the possibilities of the monomer and polymer after Dr W Chalmers from McGill University drew ICI's attention to the methacrylate polymers in 1930. Dr Crawford was working on safety glass at the time and methyl methacrylate was considered for part of its construction. In 1932 Dr Crawford discovered the now well-known method of making the monomer. The process used cheap and readily available starting materials, namely propanone, hydrogen cyanide, methanol and sulphuric acid. A patent covering this process was granted on 12th August 1932.[1]

The process, known as the Stevenston Process, is with some modifications still used today throughout the world for the manufacture of methyl methacrylate and related materials. The first production of cast sheet from this monomer was achieved by H J Tattersall of Nobel division in 1933 in Ardeer.

The manufacture of acrylic sheet was patented in 1934 by ICI under the trade name "Perspex", from the Latin to see through. The patent noted:- "It is a matter for surprise that the polymer of methyl methacrylate differs so sharply in physical properties from the polymers of known closely related compounds..."[2]

The first methyl methacrylate plant was built at the Cassel works in Billingham and when everybody was celebrating Guy Fawkes night in 1934, the first recorded batch of MMA was being produced. Plastics Works at Billingham was also established during 1934 to produce "Perspex" acrylic sheet and "Diakon"[3] acrylic moulding powder. This was also the year Adolf Hitler became Führer of Germany.[4]

Fig.14. Cassel works in the 1930s. Courtesy of ICI.

There is a wide gulf between the days when Perspex was produced with skill and a certain amount of luck in squares 14 x 14in. and the present day when sheets can be cast up to 3 metres by 2m. or extruded into even longer lengths. In the early days the men worked continuously in temperatures of 40°C in the casting chamber and were weighed every Friday to check on loss of weight. They also seemed to have lost their appetites except for a desire for tinned fruit.[5]

Production was unpredictable in those days when the syrup, produced in 45kg lots in large round bottomed flasks constantly agitated and warmed in hot baths, would react violently and scatter broken glass all over the room; when bits of dust from the atmosphere had to be lifted off the surface of the cast with a spatula; while perspiration was liable to drip from the foreheads of the men on to the cast and also have to be removed before the top glass could be lowered into place; and other problems of levelling the cast and preventing leaks from the moulds all had to be overcome. When the operator was satisfied the filled mould was left stacked under the benches for 24 hours to harden and hopefully a perfect sheet was produced.[6]

When the war came the tempo increased, research continued and before long the early difficulties were solved. The government stipulated where the plants had to be built as Perspex was classified an essential war material. Thus Darwen in Lancashire was chosen as one of the homes for Perspex; there were redundant cotton mills which could be converted cheaply, the almost perpetual cloud cover gave protection from any bombers and also the distance from the east coast was a factor.

In the early days, and during the war, only clear sheet was produced and all of the production went to the Triplex glass company to make safety glass. Later production was given over completely to aircraft glazing. Other articles could only be made from off-cuts or broken pieces.

During the war years Perspex had been used by the aircraft industry in large quantities in the manufacture of aircraft like the Hurricanes and Spitfires of Fighter Command, the Lancasters and Mosquitoes of Bomber Command, and the Sunderland flying boats of Coastal Command. When the war ended production of these aircraft stopped abruptly and the demand for Perspex fell away very sharply. The demand for clear (sheet) dropped by nearly half of the peak in 1944, and markets had to be found in other areas.

Within a year sales were back to the wartime peak as industry quickly recognised the virtues of Perspex. Fortunately for the Perspex business, the first trials with corrugated Perspex sheet for roof lighting had been a great success for factory and farm buildings damaged during the war. To produce corrugated sheet the operators lifted the equivalent of 80 tons of metal during a shift as the bars were all hand laid onto the heated sheet, and lifted off when the sheet was cold. Polymethyl methacrylate is little affected by sunlight or weather - a property which had been so clearly demonstrated during its wartime service in aircraft glazing, and which now stood it in good stead as a material for roof lights. Development work that had been carried out at Darwen made it possible to offer corrugated Perspex in profiles to match other corrugated roofing materials. New styles of roof lighting were now possible, styles that were both more efficient and cheaper than the old traditional systems. In all types of building, from schools to hospitals, from factories to farm buildings, architects were learning the advantages of the new material.

Coloured and translucent grades of Perspex were developed and, for a couple of years or so after the war, were used to make fancy goods, souvenirs and the like. Unfortunately in a market starved for several years of "luxury" goods, virtually any item however shoddy and ill-designed found a buyer. Perspex began to be regarded by the average person as fit only for such tawdry use. As life returned to normal the market for shoddy fancy goods began to disappear. It became obvious that if Perspex were to take its rightful place as a material, and if people were to appreciate its true merits then sound design and fabrication techniques had to be developed and promoted. Imaginative new shapes and forms for indoor lighting, particularly in public buildings, began to appear and at last it seemed that Perspex was indeed being used to some advantage - it was growing up.

Fig.15. Perspex corrugated sheet used in roof lights for Bow Street
Station locomotive shed. Courtesy of ICI.

In 1948 an Australian company had the idea of using Perspex for baths
because of the high cost of moving cast iron ones around the country. It was a
success and a photograph was taken in 1984 of the first bath installed in
Australia and was only removed for the photo. It still looked as good as new. The
idea was a great success and when the manufacture of baths made from Perspex
began to appear economically practical in the UK, there were nearly 10 years of
accumulated evidence of the highly satisfactory performance of thousands of such
baths in Australia.

In 1957-58 the first baths made from Perspex were launched onto the British
market. Made from 6ft by-4ft by 3/8th inch sheet, they were offered in 8 colours
- ICI technologists were convinced that Perspex was a serious contender in the
sanitaryware market but not everyone was persuaded at the time. First of all, sound
fabrication techniques had to be developed which would provide baths that were
strong, dependable, functional and attractive but also economical in materials,
equipment, and labour time. In addition, cradling systems for plumbing and fixing
had to be perfected, and the co-operation and approval of the trade obtained.

Concerns over properties such as scratch-resistance, heat resistance and
chemical resistance, were shown to be of no importance. But it was the beautiful
rich depth of colour and thermoforming capability of Perspex which soon
demonstrated that Perspex was the ideal material from which to make high quality
baths. In order to test Perspex, baths were installed early in hospitals and one such

bath installed in the Dermatological Dept of a major city hospital was used some 40,000 times in 8 years, equivalent to well over 100 years of use on the basis of a bath everyday. This bath was still giving satisfactory service in 1987. Today, well over 70% of all baths sold in the UK are now made from acrylic sheet, most of these being made from Perspex. Initially there were only a few colours and it was not until after 1966 when a new colouring system was put in[7], that the market for acrylic baths grew. In 1992 in the UK 75% of baths installed were acrylic.

Having developed the market in lighting the next logical step was for advertising signs and this is now (in 1993) one of the largest outlets for acrylic sheet. It resisted exposure to wind, sun and rain; dyes and colouring materials could be added to it, or painted or printed on to it; it had good dimensional stability, low water absorption and could withstand extremes of temperature, from Arctic frosts to heatwaves. In a few years Perspex signs became a familiar sight all over the world.

Fig.16. Perspex used for signs in Coventry City Arcade shopping centre - 1963. Courtesy of ICI.

Developments in coloured acrylic started with a few opals and then expanded into transparent tints and opaque colours. In 1957 there were 40 colours on the range and then one Spanish designer requested some nine special colours to be added. Today, in 1993, that colour range has extended to over 4000 colours over the years.

Also during the late 1940s, Perspex attracted the attention of sculptors because of its clarity, properties and availability in large blocks. Arthur Fleischmann, one of the best known, began carving life sized figures and using it for water sculptures. He did a lot of religious pieces and also sculpted a large DNA molecule and incorporated it into a water sculpture. This is now in the Melbourne Museum of Modern Art.

Fig.17. Lot's wife sculpted in Perspex from a solid block by Arthur Fleischmann. Courtesy of ICI.

During the sixties approximately 45% of the total Perspex production was sold overseas, and the usage of Perspex per head in the UK exceeded that of the US. If we look at the growth in production capacity of Perspex then the graph clearly illustrates the expansion rate.

GROWTH IN CAPACITY AT ICI

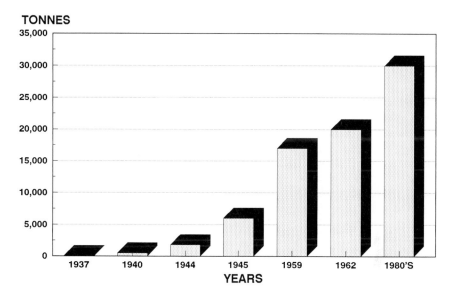

TONNES

Fig.18. Growth in Capacity at ICI

The sixties saw an expansion in the variety of applications for Perspex acrylic sheet and Diakon moulding powder. The former was used for TV implosion guards, although Diakon quickly took over on grounds of ease of manufacture by injection moulding. Windshields for motor scooters and bikes was another large outlet for sheet, together with safety goggles and shields for machinery, domelights for flat roofs and barrel vaults (the latter on the Continent in particular).

A new extruded acrylic sheet was developed, which involved putting MMA monomer in one end and getting sheet out the other end. Unfortunately the market was not ready for it and it was dropped.[8] Later in the 1970s extruded acrylic sheet did find a market but the sheet was made from polymer granules such as Diakon.

In the early days it was realised that Perspex was tolerated in the human eye and this led to its use in the medical industry not only for incubators and prosthesis but also in eye surgery. Perspex CQ is used for intraocular eye implants to restore vision in cases of cataract. Another use for CQ in the sixties came from St George's Hospital. We were asked to make a half inch thick slab in a hurry (in less than a week), for the first ever heart pacemaker to be inserted into the human body. To show how quick we had to polymerise it, half inch thick sheet in those days took 14 days.[9]

The versatility of Perspex embraces literally thousands of uses, some of them requiring thousands of tons and others but a few square inches. It is no exaggeration to say that Perspex is the most attractive of all man-made plastics, and because it can be used in so many versatile ways, it is possible to adapt it to the changing industrial scene. In 1959, an exhibition showing how Perspex can be used with effect by the designer in industry was staged at the Design Centre, London. There were sixty-seven exhibits. Opening the exhibition, Mr. Paul Reilly, Director of the Council of Industrial Design, said that if it had not been for Perspex there might not have been a Design Centre at all. This went back to 1945 or 1946, when the Chairman of ICI Plastics Division, Sir Walter Worboys, had the idea that as Perspex and other plastics were about to be "demobilised" it would be as well to look ahead, to seek the help and advice of industrial designers in considering future applications of the material. This activity commended itself to the Board of Trade. Sir Walter Worboys became a member of the Council of Industrial Design in 1947 and in 1953 was appointed Chairman of the Council by the President of the Board of Trade. In 1956 the Design Centre opened. If it had not been for the energy, drive and imagination of Sir Walter it would not have seen the light of day. "He fought the battles that won for us the Design Centre," said Mr. Reilly,"I hope this exhibition will stimulate other manufacturers of raw materials to set the sort of lead in design that   ICI   have done," he concluded.

In building, Perspex has come a long way from the first corrugated rooflights of forty years ago. It is still used for glazing rooflights but often now on a far grander scale. Throughout the world, where local building regulations permit, Perspex has been used by architects in many imaginative ways. Indeed, the Reptile House in Rome Zoo has a dome constructed entirely from Perspex with no metal supporting frame. It measures nearly 50 metres across and is over 5 metres high in the centre. It also has a tunnel across the middle through which members of the public may walk in safety.[10] Perspex has also been used for dramatic and beautiful staircases.

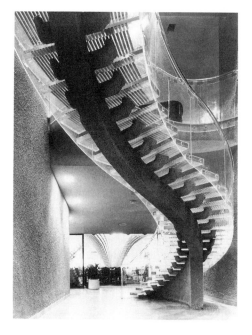

Fig.19. A staircase built in the Middle East in the 1970s. Courtesy of ICI.

Fig.20. A Perspex model of St. Paul's Cathedral built to show areas of stress and where strengthening was needed to support the dome. Courtesy of ICI.

Perspex also has important uses in the making of models - not just toys, but serious scale models of complex structures. Scale models have been used by engineers for many years to help predict the behaviour of a full scale structure or to examine the stresses in it when it is loaded. Perspex models can be constructed to show how a complex piece of machinery functions. In hospitals Perspex is used in a variety of equipment but perhaps one of its more vital properties is its transparency. A Perspex cover for an incubator is completely transparent and has no nooks and crannies to trap dirt and bacteria. Many people are alive today because they spent a few weeks in such an incubator when they were born.

Furniture made from Perspex continues to be popular due to the artistic shapes that can be produced, and in recent years Perspex has been used for sculpture and jewellery. This has established the product as an artistic material to be worked and fashioned, probably more than any other plastic material.

Special grades of Perspex were developed for specific applications. Perspex VE, for example, contains UV light absorbers that filter virtually all ultra-violet light, and this material is used as screening, glazing or display cabinets to protect priceless works of art from damage by ultra-violet light. Perspex can be made containing infra-red absorbing dyes to reduce heat build up in buildings. In 1979, ICI introduced extruded Perspex TX acrylic sheet, with improved vacuum forming characteristics and has found a place in a variety of applications, ie double-glazed caravan windows. Perspex thick sheet is used as glazing for ocean-going boats and yachts, thin sheet for secondary double-glazing and special sheets for sound insulation. Since the 1940s Perspex block of up to 100mm thickness has been available for machining prototypes and engineering applications. Perspex block can also be used for security glazing, being resistant to attack from small arms and shotguns.

Add all these products, and others, to the wide and varied applications of Perspex and one realises the great influence that John Crawford's discovery has had on our way of life today. Today acrylic sheet is freely available and, with the ambitious plans for future production now in hand,   ICI is confident of being able to satisfy all demands in the future, which looks very exciting. During the 1980s new products came on the market for both sheet and in other areas. "Asterite" acrylic casting dispersion was developed which is a highly filled composite, followed by "Avron" which is fire retardant Class 1/2 depending on thickness. In the resin field,"Modar" was developed which is a modified acrylic resin suitable for pultrusion and injection moulding.

REFERENCES
Unless otherwise indicated all the material used has been taken from references 1 and 4.

1.   25 Years of PERSPEX, ICI Plastics Division, Welwyn Garden City, 1959; BP 555,687.

2.   BP 555,687.

3.   PERSPEX and DIAKON are trade marks of ICI plc.

4.   The first Fifty Years of PERSPEX, ICI Acrylics, Darwen, Lancs., 1984.

5.   Unpublished ICI data.

6.   Taking Shape, ICI Publication, ICI Acrylics, Darwen, Lancs., 1984-92.

7.   The first Fifty Years of PERSPEX, ICI Acrylics, Darwen, Lancs., 1984.

8.   Ibid.

9.   Ibid.

10.   Taking Shape, ICI Publication, ICI Acrylics, Darwen, Lancs., 1984-92.

Fibre Reinforced Composites

Brian Parkyn

FORMERLY SCOTT BADER COMPANY LTD, WOLLASTON,
WELLINGBOROUGH, NORTHAMPTONSHIRE, UK

Importance of History

The importance of history is that it enables us to understand the times in which we live by placing them in their true perspective. The fact that history is often interesting is coincidental. Indeed, were the interesting portions of history the only ones to survive, our knowledge of the past would be confined to a disconnected series of incidents illustrating the more salacious behaviour of our forbears. We should lack detailed knowledge of the evolution of most of the factors which go to make up the complex world of today.

The beginnings of many of these factors were often quite unpretentious and their ultimate importance could not always be seen at the time. In much the same way, the study of the development of fibre reinforced composites will serve to amplify the knowledge and understanding of our present composites industry.

Definition of Fibre Reinforced Composites

That familiar verse in the 'Book of Ecclesiastes' says "There is no new thing under the sun", and this is certainly true for fibre reinforced composites. After all, wood consists of the natural resin lignin reinforced with cellulose fibres. That is why it is anisotropic. Even man-made fibre reinforced composites have their origins in pre-history, such as the Welsh coracle and the Irish currachs still used by fishermen today.

High pressure laminates of paper and phenol formaldehyde resins for electrical insulation have been made since 1913. In 1937 de Bruyne greatly increased the tensile strength of this kind of composite by using phenolic resins reinforced with long staple fibres such as silk and flax, and even considered their use for aircraft manufacture[1]. By 1939, interest moved to the development of high strength weather resistant plywood by moulding veneers bonded with "Tego Gluefilm". This was an extremely thin paper pre-impregnated with a special phenolic resin which was interleaved with the veneers, and moulded with heat and pressure. During World War Two the de Havilland Aircraft Company made the highly successful Mosquito aircraft with this type of plywood. Considerable numbers of motor torpedo boats (MTBs) were also made of plywood in a similar way.

Mention must also be made of "Hydulignum", impregnated and compressed wood, used for the manufacture of aircraft propeller blades by the Hordern-Richmond Aircraft Company. The most successful used a thermoplastic polyvinyl acetal resin, and the birchwood veneers were compressed both vertically and laterally.[2]

However, although all the above-mentioned **are** composites, as indeed is reinforced concrete, the subject of this paper concerns the family of materials that originated with glass fibre reinforced polyester (GRP). As the range of suitable matrices increased to include other polymers, as well as ceramics and metals, and the range of reinforcing fibres comprised carbon, asbestos, boron, aluminium oxide, silicon carbide, silicons, nitride and much else, this family of materials became known by 1970 as "Fibre Reinforced Composites". In general, they possess one or more of the following six characteristics:

- high tensile strength
- high modulus
- low density - hence high strength/weight ratio
- anisotropy - so that strength stiffness can be built into the material where it is most appropriate
- capable of being moulded with little or no pressure
- often capable of being cured without using heat

Glass Fibre

Glass fibre in one form or another has been known since the Egyptian XVIII Dynasty, that is about 1500 BC.[3] The earliest development of woven glass fibre appears to be 1713 when René Réaumur submitted some glass cloth to the Paris Academy of Science which had been woven by the Venetian, Carlo Riva. However, continuous filament glass fibres of sufficient fineness and consistency for reinforced plastics were not available commercially until the 1930s when they were developed by Owens Corning in the United States.

Polyester Resins

Although the first polyester resin was made in 1847 by Berzelius, polyglyceryl tartrate,[4] the first unsaturated polyester resin, glycol maleate, was made by Vorländer in 1894.[5] This was followed by the work of Smith in 1901 who reacted phthalic anhydride with glycerol to produce polyglyceryl phthalate, which led directly to the development of the alkyd resins used in the paint industry from 1913 onwards.[6]

The modern history of unsaturated polyester resins began with the publication of a patent application in 1933 by Carleton Ellis covering the reaction products of dihydric alcohols and dibasic acids and acid anhydrides for use as lacquers.[7] Ellis followed this with a further publication in 1940 showing the copolymerisation of maleic polyesters with monomeric styrene in the presence of peroxide catalysts.[8] This patent is also mainly concerned with the preparation of lacquers, and it is interesting to note that the benzoyl peroxide catalyst is referred to as a "drier". Muskat then showed that phthalic anhydride can be reacted with the maleic anhydride and either ethylene or propylene glycol so as to improve the compatibility of the final resin with styrene.[9]

This formed the basis of the Marco and Crystic unsaturated polyester resins first made in this country by Scott Bader in 1946. Shortly after this, ICI developed Nuron, a totally different type of polyester, dimethacrylate glycol phthalate which was copolymerised with n-butyl methacrylate.[10]

Epoxide resins can be also used with glass fibre for making low pressure reinforced plastics mouldings, but their use was somewhat retarded in the early days because of their higher cost and greater curing problems. Moss reacted glycerin dichlorhydrin with diphenylol propane in the presence of caustic soda and made the first epoxide resin in 1937.[11] This was followed by the work of Castan in Switzerland and others.[12]

Earliest GRP Products

The first commercial use of "low pressure resins" for reinforced plastics was in 1942 in the United States for the manufacture of glass cloth reinforced resin radomes for aircraft. This was allyl diglycol carbonate or CR39.[13] The resins were referred to as "low pressure resins" because unlike other thermosetting resins such as phenol formaldehyde, they did not cure through a condensation reaction, and therefore did not require high pressure during the moulding process.

In this country during the period 1946-1951, the aircraft industry totally dominated the market for GRP. At first this was mainly for the production of airborne radomes, generally the nose of the aircraft. The reason for this is self evident, because a radome had to be strong, low in weight, but most important of all it had to be transparent to radar. It could not therefore be made of metal, and the size and complexity of the moulding precluded the use of any other polymeric material available at the time.

In general they were made by borrowing a technique from the wood veneer mouldings industry which had been used so successfully during the war for the plywood Mosquito aircraft and the motor torpedo boats.[14] This process involved enclosing the mould and moulding in a rubber bag, and either applying a few pounds of positive fluid pressure through the bag to keep the lay-up in position and to consolidate the moulding during cure, or alternatively using vacuum. The whole assembly had to be heated to cure the resin, and cellophane had to be used on each surface of the GRP to exclude the air. This was because of "air inhibition" which will be referred to later. Many radomes were of sandwich construction consisting of three skins of GRP separated by rigid foam or honeycomb, and the thickness of these voids had to be accurately controlled so as to be appropriate for the "skip" distance of the radio frequency. The glass fibre reinforcement consisted of woven glass cloth, the only type of glass fibre reinforcement available at the time, and also the strongest.

The significance of the aircraft industry, and also the Royal Aircraft Establishment at Farnborough on the development of fibre reinforced composites cannot be over-stated. Indeed, credit must be given to the then Ministry of Supply for taking the first positive action which culminated in the production of polyester resins in Britain.[15] During the later stages of the war, resins had to be imported from the United States to meet the radar needs of our own aircraft. It was the Ministry of Supply which foresaw the desirability of having an equivalent resin, produced as far as possible from indigenous raw materials. As a result of this foresight, home produced polyester resins became commercially available in Britain during 1946 as mentioned earlier.

Other early products in this country in 1947, though not at the time commercially successful, included some 16ft seaplane floats and an experimental aircraft wing made of Durestos, a phenolic asbestos composite. Of course considerable efforts were made at the same time to try to encourage a boatbuilder to build a GRP hull, but the extreme conservatism of an industry always accustomed to using wood, dissuaded every boatbuilder who was approached from undertaking what appeared to be a pointless exercise.

However, what was historically significant was the attempt later, in 1947, in the United States to mould a 28ft boat for the US Navy. This was to be a personnel carrier and the work was carried out by the Winner Manufacturing Company at Trenton, New Jersey, using the vacuum impregnation process. In this a male and female mould were made by wet hand lay-up, and then the glass cloth laid over the inverted male mould which had a trough running around its lower edge. After positioning the female mould over the lay-up, resin was then put into the trough and sucked up by vacuum to impregnate the reinforcement, using a pipe and sightglass built into the top of the inverted female mould.[16]

Even after several attempts, considerable difficulty was experienced in getting all the reinforcement evenly impregnated due to short-circuiting by the resin, leaving large areas of dry reinforcement. Although some of the first GRP boats ever to be made were finally achieved using this system, it appeared to the author at the time that this was not a practical way to make a GRP boat. It seemed obvious that boat hulls should be made in the same way that the moulds had been made, that is by wet hand lay-up.

Air Inhibition and Cold Curing

However, two key developments over the period 1947-1951 were to totally transform the industry, and it is difficult to outline the full story of composites without some comment on my personal involvement for which the author craves indulgence.

It was becoming increasingly clear that two problems had to be resolved before the use of GRP could be widely extended beyond the aircraft industry. The first of these was "air inhibition". Even after the polyester resin had fully cured the surface in contact with the air remained permanently tacky. This necessitated the use of cellophane in order to exclude the air, and made the moulding of large or complex shapes extremely difficult. The second problem was that the resin needed to be cured at temperatures of at least 120°C and this, again, presented problems with large or complex mouldings.

The problem of air inhibition seemed at the time to be the easier to resolve, and so work was immediately started in the laboratory. First of all, resin films on metal plates were heat cured in the usual way with benzoyl peroxide as the catalyst, but firstly in an atmosphere of carbon dioxide, and then in an atmosphere of nitrogen. They remained tacky.

Further experiments were designed in which small amounts of wax were added to the resin in the hope that sufficient would rise to the surface at the point of cure, thus leaving a hard tack-free surface. Many waxes of varying melting points were used including beeswax, carnauba wax and paraffin wax. These were added to the resin in amounts varying from 0.1% upwards. They all remained tacky.

In 1947 the true nature of air inhibition was not understood, but believed to be analogous to paint drying, and therefore oxidation. This was hardly surprising since the reason for the development of polyester resins in the 1920s and 1930s was the attempt to make surface coatings superior to those made with alkyd resins. At the time, Scott Bader also manufactured a range of naphthenate paint driers, and by coincidence in the next laboratory work was in progress to accelerate the drying of alkyd paints by the addition of small quantities of benzoyl peroxide to the naphthenate driers. In view of this, small quantities of naphthenate driers were added to the polyester resin catalysed with benzoyl peroxide in the hope that this would resolve the problem of air inhibition. The tackiness remained.

By pure chance, a report was noticed in the American journal 'Modern Plastics' stating that a new polyester catalyst, cyclohexanone peroxide, had recently been developed by the Union Bay State Company in the United States. A sample was obtained for examination, and although the resin had to be heat cured as with benzoyl peroxide, it did show certain advantages.

Trials were then carried out with cyclohexanone peroxide, using small additions of various driers, including lead, zinc, manganese and cobalt naphthenate. When the cobalt naphthenate was used the resin set solid in the test-tube in a few minutes, before there was time to apply it to a panel. The resin, however, remained tacky. So although a method of cold curing polyester resins had been accidentally discovered, the problem of air inhibition remained. Indeed, it was not solved until 1951.

By a remarkable coincidence, another cold setting system was stumbled upon at almost the same time. It was found that dimethyl aniline reacted with explosive violence with benzoyl peroxide, so that polyester resins could be cold cured by using benzoyl peroxide. A patent was immediately applied for,[17] but after it was found to be already in use in the United States later that year the patent was subsequently dropped. This system was not as effective as cyclohexanone peroxide and cobalt naphthenate for two reasons. Firstly, it discoloured the resin, which made it unsuitable for GRP roof sheeting which had just started to be manufactured. Secondly, as larger mouldings came to be made it was not possible to give the resin a long "green life" without the risk of serious undercure.

However, having succeeded by accident in developing the first cold setting polyester resin system, the work had to continue to try to find a solution to the other major problem, air inhibition. Since all the experiments so far had been unsuccessful, including the addition of wax, it was decided to examine the chemistry of the unsaturated polyester resin.

At that time the principal commercial polyester resin was made by the polyesterification in an inert gas of equal molecular proportions of maleic anhydride and phthalic anhydride with a molar excess of ethylene glycol of about 25%. At a certain point during the reaction, measured by the fall in the acid number of the resin, acetic anhydride was added to remove the excess free hydroxyl groups from the remaining glycol. This was then distilled off under reduced pressure. About 0.02% t-butyl catechol and hydroquinone was then added as a stabiliser, and finally about 15% of monomeric styrene.

First of all the styrene was replaced by other suitable monomers, including di-allyl phthalate and di-vinyl benzene. Then the maleic anhydride was replaced by fumaric acid. Other glycols were used including tri-ethylene glycol and propylene glycol. In the latter case it was found that by using propylene glycol and adjusting the molar proportions so that the acetylation process could be omitted, a resin was produced that at first appeared to be non-air inhibited. This was in 1948 and the resin was marketed as "the first non-air inhibited resin". Unfortunately, in practice, although the resin was less air inhibited than earlier resins, it was still not ideal, and the premature claim for it had to be withdrawn. Finally an air-drying polyester resin was made based on tetrahydrophthalic anhydride, fumaric acid and diethylene glycol as described in a British Patent as early as 1944.[18] However the resin was difficult to cure fully, and it was too flexible for use·in glass fibre reinforced mouldings.

The usual practice in the laboratory at the time was to mix polyester resin in small aluminium containers. These often had to be discarded after use and they were therefore expensive. In 1951 supplies of small waxed paper cartons became available, and being far cheaper, they came into general use. On the twelfth of March that year some resin which had been mixed and cold cured in one of these paper cartons was found to have a totally hard tack-free surface. Work was immediately restarted with wax additives, but instead of using quantities of paraffin wax from 0.1% upwards as on the earlier occasion, quantities were reduced from 0.1% downwards. Very quickly it was established that the exact quantity was extremely critical, and with 0.03% of paraffin wax of the appropriate melting point, totally non-air inhibited polyester resins could be produced.[19]

Diversification of Industrial Uses

The resolution of the twin problems of cold curing and air inhibition led in 1951 to a dramatic diversification of the uses of GRP outside the aircraft industry. Already in 1948 with the introduction by Fibreglass of a random glass mat, a far less expensive and more versatile form of reinforcement was available. Although mouldings made with glass mat lacked the very high tensile strength of woven glass cloth required by the aircraft industry, it was suitable for a very wide range of applications. Indeed the first continuous production of translucent corrugated roof sheeting using glass mat became a major outlet from 1948 onwards. Aircraft manufacturers were also starting to use GRP for many other kinds of application in addition to radomes, These included at first window surrounds, air ducting, bulkheads and wing tips.

In 1951, shortly after the resolution of the problem of air inhibition, an event took place that was to lead to a train of happenings which would transform the entire composites industry. A telephone call was received from a Geoffrey Lord who said: "I believe in fast cars, fast women and fast living. I want to make a 16ft plastics boat and I believe that you can help me". Lord was an ex-naval officer, and also a marine architect and engineer, and with two other ex-naval officers, Petrie and Bevan, he had established a small boatyard in Blyth, Northumberland, called the North East Coast Yacht Building & Engineering Company. His aim, unlike the traditional boatbuilders encountered in 1947, was to apply modern production techniques to moulded plywood boat construction. He now decided to turn to GRP.

Within three weeks the first wet hand lay-up boat had been made using a plywood female mould. It was named "Wildfire". After fitting out the 16ft boat with a centre-board, rudder and sails, Lord, who was an experienced sailor, went on to win every sailing dinghy race in the country. So successful was he that the Royal Yachting Association had to consider making special rules for GRP boats. Suddenly the entire yachting world was buzzing with the success of "Wildfire". GRP had arrived.

Another extraordinary event that did so much for the composites industry involved an ex-racing driver called Spike Rhiando. In 1952 he built himself a GRP motor scooter and decided to attempt to break the London to Capetown land speed record. Unfortunately after fourteen days he broke down in the middle of the Sahara Desert, and was only rescued in the nick of time by a party of French geologists.[20]

The real significance of this story in the history of composites is that it was published in all the national newspapers, not just the plastics press but the Daily Express, the Daily Mail and even the Financial Times. GRP was at last being talked about in Fleet Street and being drawn to the attention of the world. As already explained, all the early applications of reinforced plastics were in the aircraft industry. The scientists at RAE Farnborough, and the engineers in all the many aircraft manufacturers we had at that time, knew all about GRP and how to use it. But GRP was unknown to the wider industrial world. It was Lord, Rhiando and many other rather eccentric pioneers whose exploits first brought awareness of GRP to a wider world.

GRP started to be talked about by car manufacturers, serious boatbuilders, the Royal Navy, and by the manufacturers of chemical storage tanks and road tankers and architects and so on. This transformation of the composites industry from aeronautics to industry at large took place just about 40 years ago.

Growth

In 1953 Halmatic of Havant built "Perpetua", at 45ft the largest composites boat ever made up to that time. From then onwards this country has led the world in building ever greater boats, culminating in the first all composites minesweeper HMS Wilton in 1972. Built by Vosper Thorneycroft for the Royal Navy, it has an overall length of 153ft and weighs some 450 tons. This was followed by a new class of even larger mine counter-measure vessels some 200ft in length and weighing 625 tons.

1955 saw the earliest GRP car bodies, lorry cabs and the mass production of motorcycle fairings. Caravans soon followed, as well as milkfloats, ice cream vans and ambulances. British Rail became interested and started moulding GRP canopy ends for their new diesel railcars. After this they started moulding front-ends of locomotives, door surrounds and then doors, and finally large numbers of press moulded passenger seats.[21]

Lloyds set up a reinforced plastics technical committee to establish standards for ships' lifeboats, which by 1961 were being made in considerable quantity. The RNLI soon followed and the first shore-based lifeboats were approved. Subsequently many 54ft Arun Class lifeboats were made, not only for use by the RNLI but also for authorities world-wide. The Trinity House Agency was also supplied with GRP vessels.

The first highly engineered filament wound pipes were made in 1957, to be followed by pressure vessels and parts of chemical plant. Liquid storage vessels and silos over 100ft high were made, road tankers, buses, box vans, deckhousing for trawlers and large sea buoys. In the early days the use of GRP in buildings was non-load-bearing, consisting mainly of roof sheeting and curtain wall cladding. From 1959 onwards, GRP was being used structurally for building self-supporting roofs for sports stadia and swimming pools. The first fleet of 100ft purse seine trawlers were made in Peru using British materials and know-how, and it began to be used for making moulds for reinforced concrete. Finally structural bridges were made entirely of GRP, the largest up to date being the bridge over the River Tay at Aberfeldy completed in 1992. This has a central span of some 200ft with poltruded box-section towers 60ft high, and has an overall length of about 400ft. The list of significant applications outside the aircraft industry is now almost endless in its diversity.

But the aircraft and defence industries continued to extend their use of composites as well, some at the very limit of composites technology. Radomes became larger and more complex. Large ground-based radomes were made. Increasing numbers of parts of aircraft structures and fittings, both military and civil, used composites for its high strength and weight saving capability. The well publicised American Stealth aircraft was only possible because of carbon fibre reinforced composites. Ballistic missiles and other weapons, parts of submarines and much else, all depend on composites.

Quality Control

Although from the beginning the use of composites on aircraft was subject to rigorous control by the Air Inspection Directorate (AID) for military aircraft, and the Air Registration Board (ARB) for civil aircraft, the very nature of fibre reinforced composites leads to considerable quality control problems.

In the United States the Reinforced Plastics Section of the Society of the Plastics Industry (SPI) was set up as early as 1945, not only to disseminate knowledge, but to look at standards. In this country a Reinforced Plastics Technical Committee was established by the British Plastics Federation in 1952, leading in 1957 to the foundation of the Reinforced Plastics Group, now the Composites Group.

But the main thrust for quality control came from the defence research establishments. This included in particular the Royal Aircraft Establishment (RAE), the Telecommunications Research Establishment at Malvern (TRE), and the Military Engineering Experimental Establishment at Christchurch (MEXE). It was not only a question of determining the resin/glass ratio, or the fibre orientation, but also the degree of cure of the final resin. After all with other polymers, and indeed metals, the chemical and physical properties of the material are controlled during their manufacture; with composites it is determined at the point of moulding.

This first became apparent in 1954 when the Bristol Aeroplane Company (as it then was) moulded a large experimental marine structure for MEXE. When the initial structure was floated in the River Stour at Christchurch, very rapid leaching of the resin occurred leading to extensive pinholing and loss of water-tightness. With the object of establishing satisfactory testing techniques for composites structures, the Joint Services Research and Development Committee on Plastics set up a Panel (Sub-Committee 2) to investigate the problem of the degree of cure of polyester resins. The work of Sub-Committee 2 from 1954 to 1959 was seminal in changing fibre reinforced composites from a casual backyard industry to a more exact science.[22]

Select Committee Enquiry

No account of the history of fibre reinforced composites would be complete without reference to the enquiry of the Select Committee on Science and Technology on Carbon Fibres chaired by the author during the 1968-1969 session of Parliament.[23]

Carbon fibre has been known for over a hundred years. The electric lamp filament developed by Joseph Swan and Thomas Edison was made by carbonising various fibre precursors including viscose, cellulose, cotton, grasses and bamboo. The exceptionally high theoretical strength and stiffness of long carbon fibre was not achieved however until a method of manufacturing a highly orientated crystalline fibre was discovered by the Royal Aircraft Establishment in 1963. The process involves the controlled carbonisation of polyacrylonitrile fibre as the precursor.[24]

Rolls Royce had already started to develop gas turbine blades from glass fibre reinforced epoxy resins in 1955. In 1965 they received some carbon fibre from RAE and quickly established that carbon fibre reinforced composites were far superior. They therefore decided to develop compressor blades made of CFRP for their new RB211 engines. In spite of its high cost, this was more than compensated by the improved aerodynamic efficiency of the blades, and therefore the power output and noise improvement of the engine.

The reasons why a Select Committee of the House should have decided to examine the somewhat esoteric subject of carbon fibre reinforced composites in so much depth are not relevant to this paper. What is relevant, is that when the Report was published in 1969 it was given quite exceptional publicity by radio, television and the press. Indeed, many newspapers, including the Financial Times, carried full-page articles relating not only to carbon fibre reinforced composites, but to the relative merits of boron fibre, sapphire and so on. For the first time the world at large became aware that composites need not be of carbon fibre with a polymer matrix, but could be of fibre reinforced metals, glass and ceramics. This Report did for the new high technology composites what the adventures of Lord, Rhiando and others did for glass fibre reinforced plastics in 1952. Indeed, it was even the subject of a debate in the House of Commons.[25]

REFERENCES

1.   N.A. de Bruyne, Journal of the Royal Aeronautical Society, 1937, 41.

2.   H.R. Fleck, 'Plastics', English Universities Press for Temple Press Ltd., London, 1943.

3.   B. Parkyn, Journal of the Royal Society of Arts, February 1963, CXI, 205-223.

4.   B. Parkyn, L. Lamb, L & B.V. Clifton, 'Polyesters', Iliffe Books, London, 1967.

5.   D. Vorländer, Annalen, 1894, 280, 167.

6.   B. Parkyn, Transactions of the Plastics Institute, July 1952, XX, pt.41, 36-56.

7.   C. Ellis, U.S.P. 1897977, 1933.

8.   C. Ellis, U.S.P. 2195362, 1940.

9.   I. Muskat, U.S.P. 2423042, 1947.

10.   R. Hammond, B.P. 630370, 1949.

11.   W. H. Moss, B.P. 506999, 1937.

12.   Gebrüder de Trey, Swiss Patent 211116, 1938.

13.   B. Parkyn, 'Chemistry of Polyester Resins', Composites, January 1972, 29-33.

14.   B. Parkyn & G. C. Hulbert, 'Twelve years of reinforced plastics', Applied Plastics, April 1959.

15.   Ibid.

16.   B. Parkyn, 'Boat hulls by low-pressure lamination', British Plastics, March 1950.

17.   B. Parkyn & E. Bader, B. P. Application 22257, 1947.

18.   Amercan Cyanamid Company, B. P. 592046, 1944.

19.   B.Parkyn & E. Bader, B. P. 713332, 1954.

20.   Daily Express, 9 March 1953.

21.   B. Parkyn, 'Glass Reinforced Plastics', Iliffe Books, London, 1970.

22.   'The assessment of cure of resins', Report of Sub-Committee 2, Reinforced Plastics, January 1961, 102-113.

23.   'Carbon Fibres', Report from Select Committee on Science and Technology, House of Commons, February 1969.

24.   W. Johnson, L.N. Phillips, W. Watt, B. P. 1110791, 1968.

25.   B. Parkyn, Parliamentary Debates House of Commons, 21 July 1969, 787, column 1332.

Subject Index